GEOTECHNICAL SPECIAL PUBLICATION NO. 145

SEISMIC PERFORMANCE AND SIMULATION OF PILE FOUNDATIONS IN LIQUEFIED AND LATERALLY SPREADING GROUND

PROCEEDINGS OF A WORKSHOP

March 16–18, 2005
University of California
Davis, California

SPONSORED BY
Pacific Earthquake Engineering Research Center, University of
California at Berkeley
Center for Urban Earthquake Engineering, Tokyo Institute of
Technology
The Geo-Institute of the American Society of Civil Engineers

EDITED BY
Ross W. Boulanger
Kohji Tokimatsu

Published by the American Society of Civil Engineers

Cataloging-in-Publication Data on file with the Library of Congress.

American Society of Civil Engineers
1801 Alexander Bell Drive
Reston, Virginia, 20191-4400

www.pubs.asce.org

Geotechnical Special Publications

Preface

The last decade has produced significant advances in our understanding of the seismic performance of pile foundations in liquefied and laterally spreading ground. Accordingly, it was an opportune time to bring together researchers and practitioners to exchange and document their most recent findings.

The workshop on "Seismic performance and simulation of pile foundations in liquefied and laterally spreading ground" was held in Davis, California, on March 16-18, 2005. This workshop was hosted by the Pacific Earthquake Engineering Research (PEER) Center through the Earthquake Engineering Research Centers Program of the National Science Foundation (under contract 2312001), and the Center for Urban Earthquake Engineering (CUEE) at the Tokyo Institute of Technology, Japan, through the 21st Century COE program of Japanese Ministry of Education, Culture, Sport, and Technology (MEXT). This workshop built upon two prior US-Japan workshops regarding the behavior of pile foundations in liquefied soils that were held in September 1998 and October 2002 in Tokyo, Japan.

This Proceedings was assembled under the sponsorship of the Earthquake Engineering and Soil Dynamics Committee of the Geo-Institute, and contains 25 papers from the 34 technical presentations that were given at the workshop. In accordance with the standards of practice of the Geo-Institute, each paper underwent peer review and had to receive two positive reviews to be accepted for publication. All papers are eligible for discussion in the ASCE Journal of Geotechnical and Geoenvironmental Engineering and for ASCE awards.

A total of 47 participants from industry and academia attended the workshop, including individuals from Japan, England, Taiwan, and the US. The workshop participants were:

Akio Abe	Stuart K. Haigh	Kyle M. Rollins
Tarek Abdoun	Ching-Lung Hung	Masayoshi Sato
Abbas Abghari	Susumu Iai	Tom Shantz
Scott A. Ashford	Jun Izawa	Hiroko Suzuki
Victoria Bennett	Jonathan Knappett	Akihiro Takahashi
Ross W. Boulanger	Kohji Koyamada	Jiro Takemura
Scott J. Brandenberg	Steve L. Kramer	Shuji Tamura
Jonathon D. Bray	Bruce L. Kutter	Tetsuo Tobita
Dongdong Chang	I. Po Lam	Kohji Tokimatsu
Misko Cubrinovski	San-Shyan Lin	Javier Ubilla
Stephen E. Dickenson	S. P. Gopal Madabhushi	Akihiko Uchida
Ricardo Dobry	Geoff Martin	Ryosuke Uzuoka
Ahmed Elgamal	Claudia Medina	Dan W. Wilson
Travis M. Gerber	Shinichiro Mori	Mourad Zeghal
Unit Gulerce	Yoshi Moriwaki	Jin-xing Zha
David Ha	Mitsu Okamura	

An important observation made by participants and discussion moderators was that relatively consistent viewpoints had emerged on several topics that had been strongly debated at past workshops, and that more complex questions of analysis and design were now moving to the forefront of our discussions. For example, the fundamental mechanisms of lateral subgrade reaction behavior between piles and liquefied soil were often a major point of debate in previous meetings, which of course hampered subsequent discussions related to analysis and design. At the present workshop, there was relatively consistent agreement on the general mechanisms of soil-pile interaction in liquefied soils, which can be attributed to the broad range of numerical and physical modeling studies that have systematically addressed the questions raised at previous meetings. The notable consequence of this and other areas of common agreement was more effective communication among participants when addressing complex issues of numerical simulation and design. In this regard, it is clear that the interactions and discussions among participants at this and previous workshops have helped speed advances in this challenging area of research.

We are confident that ongoing research will systematically address the numerous remaining issues that were identified and debated at the workshop, and that we will continue to see rapid improvements in our abilities for simulation and performance-based design of pile foundations subject to liquefaction hazards. Rapid progress requires the talents of many individuals, and for this reason it is very promising to see the numerous collaborations that have been established among different research groups in the past several years.

In closing, we would like to express our appreciation to the many individuals and organizations that made this workshop a success. This includes all the workshop participants and contributing authors for their valuable contributions to the workshop and Proceedings, the PEER Center and CUEE for their financial support of the workshop, the Geo-Institute for supporting publication of this Proceedings, and Dr. Po Lam of Earth Mechanics Inc. and Dr. Yoshi Moriwaki of GeoPentec Inc. for their sponsorship of the workshop social activities. Lastly, we especially thank Ms. Pam Pickering for her expert assistance in handling all the workshop logistics.

Ross W. Boulanger
University of California at Davis

Kohji Tokimatsu
Tokyo Institute of Technology

Contents

FACTOR AFFECTING HORIZONTAL SUBGRADE REACTION OF PILES DURING SOIL LIQUEFACTION AND LATERAL SPREADING

Hiroko Suzuki[1], Kohji Tokimatsu, Member, ASCE[1], Masayoshi Sato[2] and Akio Abe[3]

ABSTRACT

The mechanism of subgrade reaction development in liquefied and laterally spreading ground is discussed through large shaking table tests. With increasing relative displacement, the extension stress state develops on one side of a pile with significant pore water pressure reduction, while the compression stress state develops on the other side with insignificant pore water pressure reduction. The subgrade reaction is induced by the difference in stress sates in both sides, i.e., the pile is pulled by the soil on the extension side. In liquefied level ground, a pile is pulled by the soil on right and left sides alternately because the extension and compression stress states alternately develop on both sides. In laterally spreading ground, a pile is pulled only by the downstream soil because the extension stress state develops on the downstream side of the pile only when the ground moves downstream. The subgrade reaction in the laterally spreading ground consists of cyclic as well as permanent components, of which contribution depends on pile stiffness. The cyclic one with pore water pressure reduction, induced by the cyclic ground deformation, becomes large in stiff piles, while the permanent one, induced by the permanent ground deformation, becomes large in flexible piles.

INTRODUCTION

The Hyogoken-Nambu earthquake (M=7.2), that occurred on January 17, 1995, induced geotechnical problems on various structures. In particular, many buildings

---

[1] Tokyo Institute of Technology, Japan
[2] National Institute for Earth Science and Disaster Prevention, Japan
[3] Tokyo Soil Research Co., Japan

supported on piles in liquefied and laterally spreading areas settled and/or tilted without significant damage to their superstructures (BTL Committee, 1998). Subsequent studies have shown that kinematic force on piles that were arisen from large cyclic and/or permanent ground displacement was the major cause of the distress.

In order to clarify the abovementioned kinematic effects and to take them into account in seismic design, many studies on seismic behavior of pile foundations in liquefied and laterally spreading ground have been conducted with large shaking table tests and centrifuge model tests (e.g., Abdoun et al., 2003; Boulanger et al., 2003; Tokimatsu et al., 2004, 2005). It has been shown that the relationship of horizontal subgrade reaction of a pile with its displacement relative to soil (p-y behavior) controls kinematic force acting on piles and strongly depends on the pore water pressure response around the pile (e.g., Tokimatsu and Suzuki, 2004; Wilson et al., 2000); however, the mechanism of subgrade reaction development during lateral spreading as well as soil liquefaction is still unclear.

The objective of this study is to investigate and to draw a general view of the effects of pore water pressure around a pile on p-y behavior during both liquefaction and lateral spreading, based on large shaking table tests in which many pore water pressure transducers and earth pressure transducers are installed on and around piles.

LARGE SHAKING TABLE TESTS

To qualitatively investigate the mechanism of subgrade reaction development in liquefied and laterally spreading ground, the results of two shaking table tests conducted on soil-pile structure models using the large shaking table facility at the National Research Institute for Earth Science and Disaster Prevention are examined. Fig. 1 shows the test models, one with a level ground on a horizontally laminated shear box (Tokimatsu and Suzuki, 2004) and the other with an inclined ground on an inclined laminar shear box having a slope angle of 2 degrees. The dimensions of the laminar shear box were 5.0-5.5 m in height, 12.0 m in width and 3.5 m in length.

A soil profile prepared for the level ground consisted of three layers including a top non-liquefiable sand layer, a liquefiable saturated sand layer and an underlying dense sand layer, while that for the inclined ground consisted of two layers including a top non-liquefiable sand layer and an underlying liquefiable saturated sand layer. The sand used for both the test models was Kasumigaura Sand ($e_{max} = 0.961$, $e_{min} = 0.570$, $D_{50} = 0.31$ mm, $F_c = 5.4$ %). The bottom non-liquefiable layer of the level ground was compacted to form the specified relative density of 80%. The laminar shear box was then filled with water to a certain level and wet sand for the liquefiable layer was pluviated into the water. The dry sand for the top layer above the water table was air-pluviated.

A 2x2 pile group of which heads were fixed to a foundation (20.6 kN) with a superstructure (139.3 kN), was used in the level ground. Each pile had a diameter of about 320 mm and a wall thickness of 6 mm. In contrast, two single piles having a free rotational were used in the inclined ground. Both piles had the same diameter

Table 1: Test series

	Soil model		
	Level ground	Inclined ground	
Pile model	2x2 pile group (D=318.5mm, t=6mm)	Stiff pile (D=318.5mm, t=6mm)	Flexible pile (D=318.5mm, t=3mm)
Input motion	Rinkai 2.0m/s^2	Sine wave (2Hz) 2.0m/s^2	

D: Diameter of pile, t: Thickness of pile

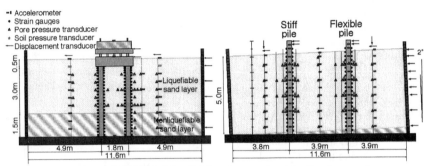

Figure 1: Test models with level ground and inclined ground

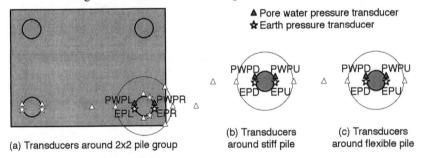

Figure 2: Transducers on pile surface

of about 320 mm but with different wall thickness of 3 or 6 mm, so the upstream pile is hereby called the flexible pile and the downstream pile the stiff pile. All the piles in both level ground and inclined ground were fixed to the base of the laminar shear box.

The test models were densely instrumented with accelerometers, displacement transducers, strain gauges, pore water pressure transducers, and earth pressure transducers. To investigate stress states in soil around a pile during soil liquefaction and lateral spreading, many pore pressure transducers and/or some earth pressure sensors were densely installed on and around a pile at a depth of about 3 m (2.5m below pile head in the level ground or 3.5m below pile head in the inclined ground), as shown in Fig. 2.

An artificial ground motion called Rinkai, produced as a design earthquake to be expected in the Southern Kanto district in Japan, was used as an input base acceleration to the test with the level ground, while a sine wave with a frequency of 2 Hz was used for the test with the inclined ground.   The maximum acceleration used for both tests was adjusted to 2.0 m/s^2.

EFFECTS OF STRESS STATES IN SOIL AROUND PILE ON P-Y BEHAVIOR IN LIQUEFIED LEVEL GROUND

Test Results

To evaluate the mechanism of subgrade reaction development in liquefied level ground, the subgrade reaction of a pile is calculated from the double differentiation of observed bending moment with depth and the displacements of the ground and pile from the double integration of their observed accelerations.   Fig. 3 shows the time histories of the followings at a depth of 3 m in the tests with the level ground:
• Displacements of soil and pile
• Relative displacement between soil and pile
• Horizontal subgrade reaction of pile
• Pore water pressures measured on both (right and left) sides of pile
• Input motion

Liquefaction develops within 10-20 s after the pore water pressure reaches the initial effective stress of about 30 kPa (Fig. 3(e)(f)). Despite reduction in effective stress, the subgrade reaction after soil liquefaction does not decrease significantly (Fig. 3(d)).   This is probably due to the reduction in excess pore pressure with increasing relative displacement. It is interesting to note that subgrade reaction increases in accordance with the reduction in pore water as well as the increase in relative displacement (Fig. 3(d)(e)(f)).

Effects of Pore Pressure Variation on p-y Behavior

To investigate the effects of pore water pressure on p-y behavior, Fig.

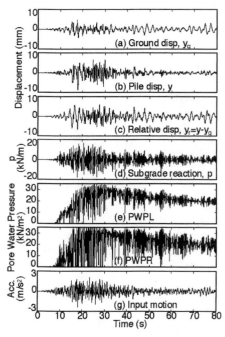

Figure 3: Time histories of displacements, subgrade reaction and pore water pressures at 3 m depth in test with level ground (Tokimatsu and Suzuki, 2004)

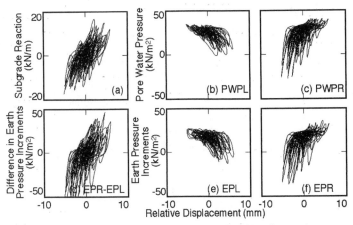

Figure 4: Relations of relative displacement with subgrade reaction, pore water pressures and earth pressure increments at 3 m depth in test with level ground (Tokimatsu and Suzuki, 2005)

4 shows the relation of relative displacement with subgrade reaction and the pore water pressures and earth pressure increments measured on both sides of the pile during liquefaction. The positive relative displacement in the figures indicates that the pile pushes the soil on the right or the soil pushes the pile from the right. The negative relative displacement, conversely, indicates that the pile pushes the soil on the left or soil pushes the pile from the left.

When the positive relative displacement develops, the subgrade reaction increases sharply (Fig. 4(a)). At this stage, the pore water pressure on the left side of the pile decreases significantly, whereas that on the right side maintains almost constant (Fig. 4(b)(c)). When the negative relative displacement develops, the pore water pressures on both sides are reversed. The earth pressure increments measured on both sides of the pile show a similar trend to that of the pore water pressures on the same sides (Fig. 4(e)(f)). The difference in the two earth pressures on both sides of the pile with relative displacement, shown in Fig. 4(d), shows the same trend as the subgrade reaction shown in Fig. 4(a). This indicates that the subgrade reaction is induced by the pore water pressure changes around the pile.

Fig. 5 shows a schematic diagram indicating how the subgrade reaction of a pile develops during

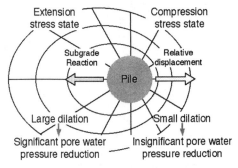

Figure 5: Stress states in soil around pile in liquefied ground

soil liquefaction. When the pile pushes the soil on the right or the soil pushes the pile from the right, the compression stress state develops on the right and the extension stress state develops on the left of the pile. On the extension side, the pore water pressure reduction becomes pronounced due to the combined effects of decrease in normal stress and soil dilation induced by the shear stress. On the compression side, in contrast, the pore water pressure reduction becomes small due to the adverse effects of increase in normal stress and soil dilation induced by the shear stress. As a result, the pile is pulled back by the soil on the extension side. Such mechanism of p-y behavior in liquefied soil is different from that in dry sand where horizontal subgrade reaction is induced by the increase in soil pressure on the compression side of the pile.

EFFECTS OF STRESS STATES IN SOIL AROUND PILE ON P-Y BEHAVIOR IN LATERALLY SPREADING GROUND

Test Results

To evaluate the mechanism of subgrade reaction development in laterally spreading ground, the subgrade reaction of a pile is calculated from the double differentiation of observed bending moment with depth and the displacement of the pile from the double integration of observed bending moment. The displacement of the ground is assumed to be equal to the observed displacement of the laminar shear box at the same depth. Fig. 6 shows the time histories for the stiff pile at 3 m depth in the test with inclined ground for the following:
• Displacements of soil and pile
• Relative displacement between soil and pile
• Horizontal subgrade reaction of pile
• Pore water pressures measured on both (right and left) sides of pile
• Input motion

The pore water pressure reaches the initial effective stress in several cycles. The ground displacement increases downstream (on the negative side) with cyclic fluctuation. It is interesting to note that the subgrade reaction develops on both positive and negative sides before liquefaction but it develops only on

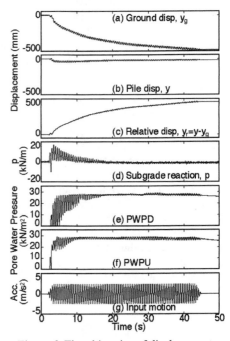

Figure 6: Time histories of displacements, subgrade reaction and pore water pressures at 3 m depth in test with inclined ground

the positive side after liquefaction. The pore water pressures after liquefaction decrease cyclically particularly on the downstream side, which could affect the subgrade reaction development, as in the case of level ground.

Effects of Pore Water Pressure Variation Around Pile on p-y Behavior

To estimate the effects of pore water pressure variation around the pile, Fig. 7 shows the relation of relative displacement with subgrade reaction and the pore water pressures and earth pressure increments observed on both sides of the stiff pile. Black lines in the figures indicate that the ground and pile move downstream and gray lines indicate that they move upstream.   When the pile and ground move downstream, shown in black lines, the pore water pressure and earth pressure increment on the downstream side of the pile decrease significantly with those on the upstream side almost constant (Fig. 7(b)(c)(e)(f)).   At this stage, the subgrade reaction becomes large (Fig. 7(a)).   When the pile and ground move upstream, shown in gray lines, the pore water pressures and earth pressure increments on both sides of the pile increase or maintain almost constant (Fig. 7(b)(c)(e)(f)).   At this stage, the subgrade reaction decreases (Fig. 7(a)).   The difference in two earth pressure increments shows good agreement in trend with the subgrade reaction (Fig. 7(d)).   This indicates that the subgrade reaction in laterally spreading ground is also induced by the pore water pressure changes in soil around a pile.

Fig. 8 shows a schematic diagram indicating how the subgrade reaction of a pile develops during lateral spreading.   When the pile and ground move downstream, the compression stress state develops on the upstream side of the pile with insignificant pore water pressure reduction and the extension stress state develops on the downstream side with significant pore water pressure reduction.   As a result, the

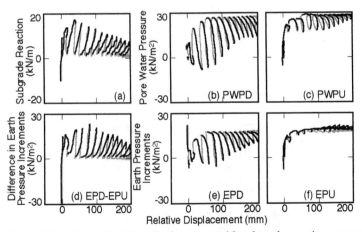

Figure 7 Relations of relative displacement with subgrade reaction, pore water pressures and earth pressure increments for stiff pile at 3.0 m depth in test with inclined ground

pile is pulled by the soil on the downstream side.   When the pile and ground move upstream, the relative displacement does not increase but decrease due to the accumulated downstream ground displacement.   Therefore, the stress states developed on both sides of the pile are considered to be unloading, leading to a decrease in subgrade reaction.   This indicates that the subgrade reaction development in laterally spreading ground is caused by the difference in stress states on both sides of a pile, as in the case of level ground.   The pile in laterally spreading ground, however, is pulled only by the downstream soil, which is different from that in liquefied level ground where the pile is pulled by the soil on both sides alternately.

Effects of Pile Stiffness on p-y Behavior

Fig. 9 shows the relations of relative displacement with subgrade reaction and pore water pressures on both sides for the flexible pile.   A comparison of trends between the stiff and flexible piles shows that a decrease in pore water pressure on downstream side as well as an increase in subgrade reaction with increasing relative displacement is more significant in the test with the stiff pile than with the flexible pile (Figs. 7 and 9(a)(b)(c)).   It is also interesting to note that the value of subgrade reaction of the stiff pile is back to almost zero at every cycle, while that of the flexible pile is not (Figs. 7 and 9(a)).   This indicates that the subgrade reaction

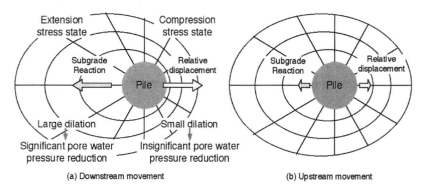

(a) Downstream movement   (b) Upstream movement

Figure 8 Stress states in soil around pile in laterally spreading ground

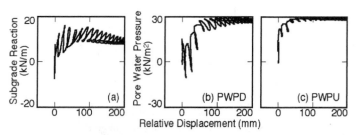

Figure 9 Relations of relative displacement with subgrade reaction and pore water pressures for flexible pile at 3.0 m depth in test with inclined ground

consists of cyclic and permanent components. The cyclic component, which might have been induced by the cyclic ground deformation with pore water pressure reduction, is larger in the stiff pile than in the flexible pile. The permanent component, which might have been induced by the permanent ground deformation, on the other hand, is larger in the flexible pile than in the stiff pile. The difference in subgrade reactions between stiff and flexible piles is probably induced by the difference in behavior between the two piles. The stiff pile, that can resist ground movement more than the flexible pile, yields larger relative displacement but smaller permanent displacement. The flexible pile, on the other hand, that can follow ground movement, yields smaller relative displacement but larger permanent displacement. As a result, the cyclic subgrade reaction becomes larger in the stiff pile but the permanent one becomes larger in the flexible pile.

CONCLUSIONS

Effects of stress states in soil around piles on subgrade reaction development in liquefied and laterally spreading ground have been investigated through large shaking table tests with both level ground and inclined ground. Discussions on the test results have shown the following:

1) In both liquefied level ground and laterally spreading ground, the extension and compression stress states develop on rear and front sides of a pile with increasing relative displacement between soil and pile. The pore water pressure on the extension side decreases due to the combined effects of extension and shear stresses, while that on the compression side maintains almost constant due to the adverse effects of compression and shear stresses.

2) The increase in horizontal subgrade reaction of a pile in liquefied and laterally spreading ground is caused by the difference in pore water pressures on both sides of the pile. The pile may be pulled by the soil on the extension side. Such mechanism of p-y behavior in liquefied soil is different from that in dry sand where horizontal subgrade reaction is induced by the increase in soil pressure on the compression side of the pile.

3) In liquefied level ground, the extension and compression stress states alternately develop on both sides of a pile. As a result, the pile is pulled by the soil on the right and left sides alternately. In laterally spreading ground, the extension stress state develops on the downstream side of the pile only when the ground moves downstream. As a result, the pile is pulled only by the downstream soil when the ground moves downstream.

4) The subgrade reaction in laterally spreading ground consists of two components. One is induced by the cyclic ground deformation, which becomes large in a stiff pile. The other is induced by the permanent ground deformation, which becomes large in a flexible pile. This is because the stiff pile resists ground movement, while the flexible pile follows ground movement.

ACKNOWLEDGMENTS

The study described herein was made possible through a Special Project for Earthquake Disaster Mitigation in Urban Areas, supported by the Ministry of Education, Culture, Sports, Science and Technology (MEXT) and US-Japan collaboration research with National Institute for Earth Science and Disaster Prevention, Rensselaer Polytechnic Institute, University of California, San Diego, Tokyo Institute of Technology and Tokyo Soil Research Co. Ms. Nakatsuji, graduate student of Tokyo Institute of Technology, assisted in examining the results of the lateral spreading test. The authors express their sincere thanks to the above organizations and person.

REFERENCES

Abdoun, T., Dobry, R., O'Rourke, T. D. and Goh, S. H. (2003). "Pile response to lateral spreads: centrifuge modeling", *Journal of Geotechnical and Geoenvironmental Engineering*, ASCE, 129 (10), 869-878.

BTL Committee (1998). "Research Report on liquefaction and lateral spreading in the Hyogoken-Nambu earthquake" (in Japanese).

Boulanger, R. W., Kutter, B. L., Brandenberg, S. J., Singh, P. and Chang, D. (2003). "Pile foundations in liquefied and laterally spreading ground during earthquakes: Centrifuge experiments & analyses", Report No. UCD/CGM-03/01, Center for Geotechnical modeling, Department, of Civil Engineering, UC, Davis.

Tokimatsu, K. and Suzuki, H. (2004). "Pore water pressure response around pile and its effects on p-y behavior during soil liquefaction", *Soils and Foundations*, JGS, 44 (6), 101-110.

Tokimatsu, K., Suzuki, H. and Sato, M. (2005). "Effects of dynamic soil-pile-structure interaction on pile stresses", *Journal of Structural and Construction Engineering*, AIJ, 587, 125-132 (in Japanese).

Tokimatsu, K. and Suzuki, H. (2005). "Effects of pore pressure response around pile on horizontal subgrade reaction during liquefaction and lateral spreading in large shaking table tests", *Proc. of the 16th International Conference on Soil Mechanics and Geotechnical Engineering*, (Accepted).

Wilson, D. W., Boulanger, R. W. and Kutter, B. L. (2000). "Observed seismic lateral resistance of liquefying sand", *Journal of Geotechnical and Geoenvironmental Engineering*, ASCE, 126 (10), 898-906.

P-Y CURVES FOR LARGE DIAMETER SHAFTS IN LIQUEFIED SAND FROM BLAST LIQUEFACTION TESTS

Kyle M. Rollins, Member, ASCE[1], Lukas J. Hales, Member, ASCE[2], Scott A. Ashford, Member, ASCE[3], William M. Camp III Member, ASCE[4]

ABSTRACT

To aid in the design of the Ravenel Bridge in Charleston, South Carolina, lateral load tests were performed on a 2.59 m diameter drilled shaft foundation before and after blasting to induce liquefaction. Excess pore pressure ratios between 75 and 100% were induced to a depth of 13 m using explosive charges. Following blasting, the stiffness of the load-deflection curve decreased to about 15% of its pre-blast value and the reduced lateral resistance led to a 100% increase in the maximum bending moment. Back-calculated p-y curves generally had concave upward curve shapes similar to those observed in previous blast-liquefaction tests at Treasure Island. The increased resistance for the 2.59 m diameter shaft could be accounted for using a constant pile diameter correction factor of nine relative to a 0.3 m diameter pile.

INTRODUCTION

The lateral resistance of deep foundations in liquefiable sand continues to be an important design issue for bridges in seismically active areas. This is certainly true for the Arthur Ravenel Bridge spanning the Cooper River in Charleston, South Carolina. When completed in June 2005, the Ravenel Bridge will have a clear span of 471 m (1546 feet), making it the longest cable-stayed bridge in North America. Geotechnical investigations for the Ravenel Bridge determined that liquefaction could occur to a depth of 40 ft on the eastern approach in a repeat of the 1886

[1] Civ. & Environ Engrg. Dept. Engrg, Brigham Young Univ., 368 CB, Provo, UT 84602
[2] ExxonMobil URC, URC-GW3-730A, P.O. Box 2189, Houston, TX 77252-2189
[3] Struct. Engrg. Dept., Univ. of Calif.-San Diego, 9500 Gilman Dr, La Jolla, CA 92093-0085
[4] S&ME, Inc., 621 Wando Park Blvd, Mt. Pleasant, SC 29464-7937

Charleston earthquake (estimated M7.3). Based on the success of the Treasure Island Liquefaction Test (TILT) in San Francisco (Ashford et al 2004, Rollins et al 2005a, Weaver et al 2005), designers included full-scale blast liquefaction testing as part of the $4 million foundation testing program for the new bridge (Camp et al 2002). Following blast-induced liquefaction, lateral load tests were performed using both conventional hydraulic load actuators and a statnamic loading device to evaluate lateral resistance of the large diameter drilled shaft foundations anticipated for the bridge. This paper will focus on the test results obtained with the hydraulic actuators and p-y curves derived from measurements during these tests. Comparisons will also be made with results from the TILT project in an effort to evaluate the effect of pile diameter on the lateral resistance of the liquefied sand.

GEOTECHNICAL SITE CONDITIONS

The soil profile at the test site generally consisted of alluvial sands underlain by the Ashley formation of the Cooper Group at a depth of about 13 m. This formation is known locally as the Cooper Marl. Groundwater was generally located between the ground surface and a depth about of 1.5 m, depending on tidal fluctuations. The sandy sediments of the coastal plain in South Carolina are typically loose Pleistocene age materials while the Cooper Marl is an Eocene to Oligocene age marine deposit. Prior to design of the bridge, a comprehensive geotechnical investigation was carried out to define the characteristics of the subsurface materials at the site. Preliminary studies were initially performed by Parson-Brinkerhoff and more detailed investigations were subsequently performed by S&ME, Inc. Based on the test hole logs, Camp et al (2002) developed an idealized soil profile for the site with six layers. This profile, with some minor modifications, is shown in Fig. 1.

The first layer typically extends from the ground surface to a depth of 1.5 m and consists of loose, poorly graded fine sand (SP) to silty sand (SM). In some cases, sandy clay layers were interbedded in this material. The surface sand was typically underlain by a sandy clay layer 1.0 to 1.5 m thick which classified as CH material. This clay layer was very soft and had an average natural moisture content of about 106%, which is approximately the same as the liquid limit suggesting that the clay is normally consolidated. The PI was typically about 70%. The third layer was also a loose, fine sand (SP) to silty sand (SM) similar to the first layer and typically extended to a depth of 5.5 m. The fines content varied considerably with depth and from hole to hole with a range from 0.5 to 28%. The fourth layer was typically located between 5.5 and 8.5 m below the ground surface. This layer was also a sand layer but contained significantly more fines. The layer typically classified as silty sand (SM) or clayey sand (SC). The natural moisture content was 30% and the fines content varied from 15 to 24%. The fifth layer generally began at 8.8 m depth and extended to the top of the Marl. This layer contained fewer fines and generally classified as a loose to medium dense poorly graded fine sand (SP).

The Cooper Marl was encountered between 12.8 and 14 m below the ground surface and extended to a depth of 85 m, which was below the base of all the test foundations at the site. The Cooper Marl is a stiff, high plasticity calcareous silt or

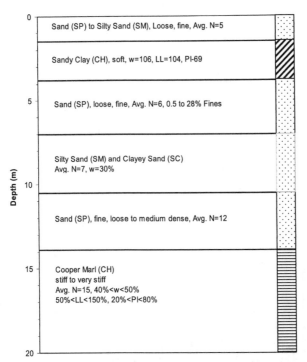

Figure 1. Idealized soil profile for the test site (Modified from Camp et al., 2002).

clay and generally classifies as MH or CH material (Camp et al, 2002) according to the Unified Soil Classification System. The liquid limit typically ranges from 50 to 90% with a plasticity index varying from 15 to 60%. The natural moisture content is typically between 40 and 60% which is somewhat higher than the plastic limit but much lower than the liquid limit indicating overconsolidation. The marl is very stiff with undrained strengths typically ranging from 100 to 200 kPa at the top of the layer and increasing with depth to a value between 200 and 300 kPa at a depth of about 45 m. Below this depth, the strength appears to remain relatively constant.

Cone penetration (CPT) soundings were performed at several locations near the test area and the results from one sounding performed by S&ME are shown in Fig. 2. The normalized cone (tip) resistance (q_{c1}), friction ratio (f_r) and pore water pressure (u) for each of the soundings are presented as a function of depth below the ground surface in Fig. 2. The q_{c1} value is given by the equation

$$q_{c1} = q_c C_Q = q_c \left[\frac{p_a}{\sigma'_{vo}} \right]^{0.5} \tag{1}$$

where σ'_{vo} is the effective vertical stress, p_a is atmospheric pressure and the adjustment factor C_Q is less than or equal to 1.7.

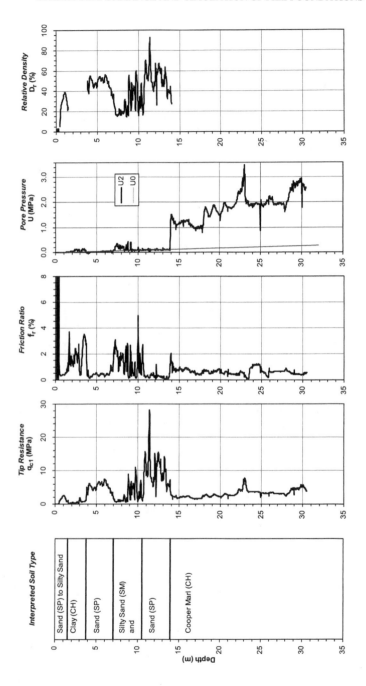

Figure 2. Interpreted soil profile and results of CPT sounding at test site.

The relative density (D_r) of the coarse-grained layers was estimated from the CPT cone resistance using the equation

$$D_r = \left[\frac{\left(\frac{q_{c1}}{P_a} \right)}{305} \right]^{0.5}$$

(2)

developed by Kulhawy and Mayne (1990) for clean sands.   The relative density profile computed using equation 2 is also plotted in Fig. 2

FOUNDATION CHARACTERISTICS AND TEST LAYOUT

Test Foundations

Test shaft MP-1 was a 2.59 m outside diameter, cast-in-steel-shell (CISS) pile.   The steel casing, with a thickness of 25.4 mm, was advanced through the sand layers and into the Cooper Marl at a depth of 16.15 m using a vibratory hammer.   The casing was then drilled out and the hole was advanced through the Marl to a depth of 47 m without casing.   The steel shell was then filled with concrete using a tremie pipe. Based on 10 tests, the concrete had an average 30-day compressive strength of 37.2 MPa (5.4 ksi).   The vertical reinforcement consisted of 36 #18 bars evenly distributed around a circle with a diameter of 2.14 m.   Confinement for the vertical steel was provided by #6 bar spirals with a pitch of 89 mm.   A 150 mm concrete cover was maintained between the spiral reinforcement and the inside of the steel case.   A profile of the shaft with two cross-sections is provided in Fig. 3.

Load Test Layout and Instrumentation

The load was applied at a height of 0.533 m above the ground surface using two MTS actuators acting in parallel, each capable of producing 2200 kN of force.   To provide a reaction, another pile with nearly identical properties was constructed approximately 8.5 m from test pile MP-1 on centers. The actuators were connected to each pile with a pinned connection to provide a free-head condition.   This connection allowed the application of cyclic compressive and tensile forces.   Each actuator was controlled with an electromechanical servo-valve and an electric hydraulic pump.

Two LVDTs were set up to measure displacements of the pile head.   They were mounted to a reference beam supported by driven piles within isolation casings.   The LVDTs were attached to MP-1 at heights of 0.533, and 1.335 m above the ground surface.   Therefore, the lower LVDT was in line with the point load application on MP-1.   A string potentiometer was connected between the reference frame and the test shaft at a height of 0.744 m above the ground surface to provide feedback for the

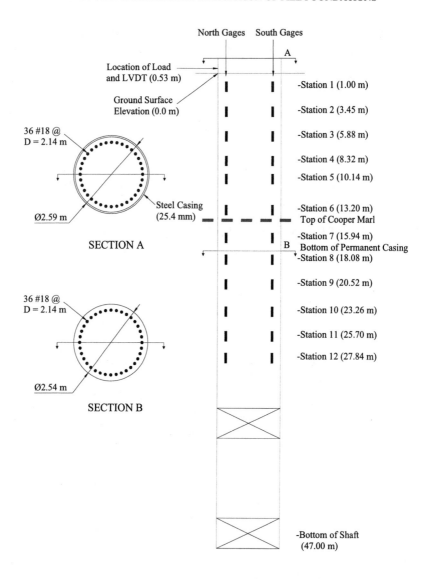

Figure 3. Profile and cross-sections through test shaft MP-3 at Mt. Pleasant test site.

servo-control system in controlling displacement. Four load cells on each of the two actuators provided a direct measurement of the applied load.

Prior to concrete placement, resistance-type strain gages were mounted on "sister bars" composed of a three foot long #4 bars and tied into the rebar cage. At each strain gage station, two strain gages were mounted on opposite sides of the pile separated by a distance of about 2.14 m. They were oriented to be in line with the direction of loading. Fig. 3 also shows the strain gage locations on pile MP-1 and their depths below ground.

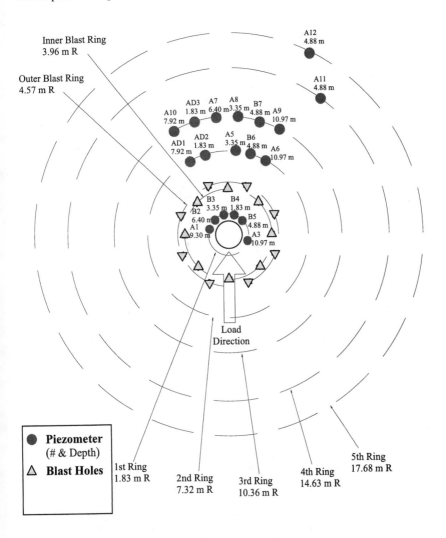

Figure 4. Layout of test pile MP-1 along with blast hole locations and piezometers.

Test pile MP-1 was also instrumented with a string of downhole electrical inclinometers to monitor slope change during loading. The inclinometers were installed inside individual guide mounts which were lowered into a grooved inclinometer casing that had been attached to the reinforcement cage of pile MP-1 and cast in place with concrete. All of the pile data (load, deflection, and strain), with the exception of the inclinometer readings, was monitored via a data acquisition system.

To quantify the build-up and dissipation of pore pressure after blasting, 19 piezometers were installed at various distances and depths around the test pile. The location and depth of each piezometer are shown in Fig. 4. Basically, three vertical arrays were installed at radii of 1.83, 7.32 and 10.36 m from the center of the test pile. All the piezometers were distributed around the front of the pile in the direction of loading to track the variation in pore water pressure with applied load. Transducers identified with a B employed piezoresistive transducers identical to those at Treasure Island (Ashford et al, 2004) while those identified with an A used electrical resistance transducers (Rollins et al 2005b). The electrical resistance transducers were more sensitive to damage during blasting and several transducers closest to the explosive charges were damaged during blasting while those further from the charges provided useful information. Additional information regarding pore pressure transducer selection, installation, and performance is provided by Rollins et al (2005b).

TESTING PROCEDURE AND TEST RESULTS

To evaluate the decrease in lateral resistance due to blast induced liquefaction, lateral load tests were performed on test pile MP-1 before and after blasting. The pre-blast load test also provided an opportunity to troubleshoot the loading system and minimize the potential for problems following blasting.

As shown in Fig. 4, two rings of explosive charges were installed around the test pile for two separate blast events. Each blast hole on the inner ring contained three 0.68 kg explosive charges centered at depths of 3.05, 6.1, and 9.15 m and a 1.36 kg charge at a depth of 12.2 m below the ground surface. The binary explosive charges consisted of ammonium nitrate and nitro-methane. Pea gravel back-fill (stemming) was placed between each charge and up to the ground surface to prevent the blast energy from simply escaping vertically. Each blast hole on the outer ring contained three 0.9 kg charges centered at depths of 4.57, 7.62, and 10.67 m. During each blast event, the charges were detonated two at a time with a delay of 250 msec between detonations. The charges were detonated beginning around the bottom ring and then moving upward around each subsequent ring to the top.

Because of the time required to apply one load cycle with the pump system, (approx. 60 to 110 sec.) it was anticipated that the second set of charges would be detonated after a few minutes of loading to keep the excess pore water pressure elevated. Unfortunately, the pump overheated in the hot, humid climate of South Carolina. Therefore, testing was limited to 5 cycles at 2700 kN followed by 5 cycles at 4400 kN for the first blast. After a delay of a few hours to repair the pump, the second set of explosives was detonated and a similar set of load cycles was applied.

After each blast, the excess pore pressure ratio (R_u) for each pore pressure transducer was computed by dividing the change in pressure by the initial vertical effective stress. Plots of the measured excess pore pressure ratio vs. depth for the three vertical arrays shortly after the first blast are plotted in Fig. 5. The R_u values are typically between 75 and 100% throughout most of the sandy soil layers. R_u values are somewhat lower at the very top and bottom of the profile.

The load vs. deflection curves before and after the two test blasts are presented in Fig. 6. As was observed in the TILT project, there is a significant decrease in stiffness after the build up of high excess pore pressure ratios (Rollins et al, 2005a). Approximately 6 to 7 times more movement is required to develop the same lateral resistance as that prior to blasting. This decrease in lateral resistance is somewhat less than that observed in the TILT experiments and may result from the fact that the percentage of lateral resistance carried by the large diameter shaft itself is much larger than that for the smaller diameter piles used in the TILT project. Because sand flowed into the space behind the pile during loading, it became necessary to apply a tensile force (negative sign) to pull the pile back to the initial position.

Both the maximum moment and the depth to the maximum moment increased substantially following blasting because of the reduced soil resistance due to pore pressure development in the loose sand. After the first blast, the maximum moment for a given load increased by about 100% in comparison with the pre-blast value. Even after some densification from the first blast, the maximum moment following the second blast was still about 70% higher than the pre-blast value. Prior to blasting, the maximum moment occurred at a depth of about 6 m; however, after liquefaction, the maximum moment occurred near the interface with the Marl at a depth of 13 m.

DEVELOPMENT OF P-Y CURVES

P-y curves were developed for three lateral load tests using procedures similar to those described by Rollins et al (2005a). Additional details are provided by Hales and Rollins (2003). The horizontal deflection (y) at a point along the length of the pile was obtained using the equation

$$y = \int \left(\int k \, dz \right) dz \qquad (3)$$

where dz is the differential element along the length of the pile and k is the pile curvature measured by the strain gages. The curvature function along the pile length was approximated using quadratic splines which matched the measured curvature at strain gage locations. Slope and deflection were taken as zero at 40 m depth. The curvature function was then double integrated to obtain deflection. At some depths where cracking occurred, curvature had to be approximated based on one strain gage with a constant adjustment factor. With minor adjustments, the computed pile head displacement was in very good agreement with the measured displacement.

The lateral soil resistance per unit length (p) was calculated by double differentiating the moment versus depth curve according to the equation

$$p = \frac{d^2 EIk(z)}{dz^2} \tag{4}$$

where E is the elastic modulus, I is the moment of inertia, and k is the curvature of the pile at depth z below the ground surface. To mitigate the effect of experimental error in the calculation of soil pressure, cubic polynomials were fit to successive sets of five nodes each and then differentiated directly at the midpoint as suggested by Matlock and Ripperger (1958) with averaging techniques developed by Gerber (2003). The back-calculated EI value for the uncracked section at the top of the pile was within about 5% of the computed value based on concrete and steel properties. Cracked EI values were estimated based on the change in moment before and after cracking. The cracked EI was typically about 55% of the uncracked EI.

Based on the back-calculated data points, a smoothed p-y curve was computed at each strain gage location and these curves are shown for a number of depths in Fig. 7. The curves tend to have a concave upward shape but are somewhat more linear than those computed for the TILT project. The p-y curves increase in stiffness as depth increases.

P-y curves were also computed at each strain gage depth using the equation

$$p = p_d A(B y)^C \tag{5}$$

proposed by Rollins et al (2005a) based on the TILT results where p is soil pressure per length (kN/m), p_d = pile diameter correction factor = 3.81ln(d)+5.6, d is pile diameter (m), $A = 3 \times 10^{-7} (z + 1)^{6.05}$, $B = 2.80 (z + 1)^{0.11}$, $C = 2.85 (z + 1)^{-0.41}$, y is horizontal deflection (mm), and z is depth (m). For the 2.59 m diameter piles at the Charleston test site, the diameter correction factor (p_d) is 9.0. The Charleston p-y curves are compared to those computed using equation 5 in Fig. 7. Despite the constant diameter correction factor, the agreement between the two curves is very good except at a depth of 1.0 m. The discrepancy at this depth is likely due to the relatively low excess pore pressure ratio (R_u=0.69) which leads to a stiffer response than that predicted by equation 5.

Using the back-calculated p-y curves for the liquefied sand and the non-linear properties for the test pile, lateral pile load analyses were performed using LPILE for comparison with measured response. Good agreement was obtained between measured and computed bending moment versus depth curves.

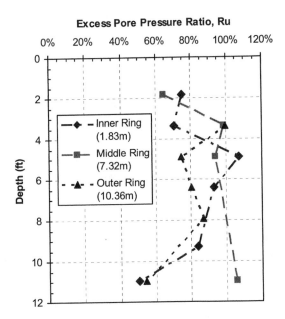

Figure 5. Excess pore pressure ratio (R_u) vs. depth for the vertical arrays at 1.83, 7.32, and 10.36 m from the center of test pile MP-1.

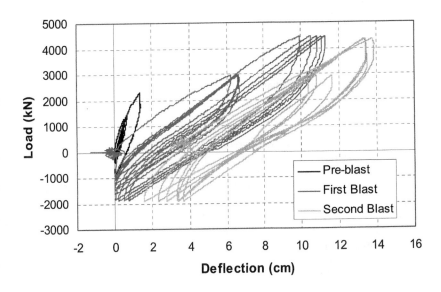

Figure 6. Pile head load vs. deflection curves for test pile MP-1 prior to blasting and for two load tests following blasting.

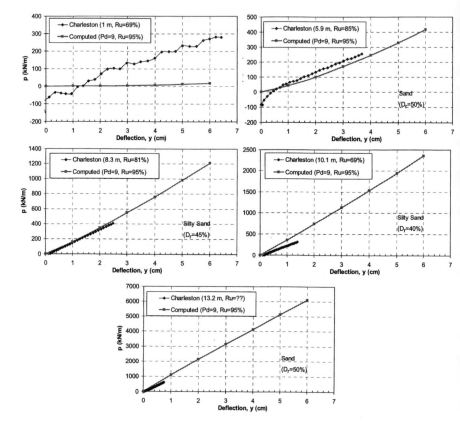

Figure 7. Comparison of back-calculated and computed p-y curves for the liquefied sand at Charleston, South Carolina test site.

CONCLUSIONS

1. The back-calculated p-y curves obtained from the lateral load tests on large diameter test piles in Charleston had shapes similar to those from the TILT experiment. Soil resistance increased with displacement and with depth.
2. The increased resistance provided by the large diameter (2.59 m) test pile could be reasonably approximated using the equations proposed by Rollins et al (2005a) for a 0.3 m diameter pile with a constant diameter correction factor of nine.
3. Full-scale blast liquefaction tests can provide useful information for design of foundations on major bridge projects.

ACKNOWLEDGMENTS

Funding for the full-scale load tests was provided by the South Carolina Department of Transportation and Modern Continental South was the contractor for the testing program. Applied Foundation Testing installed the strain gages, piezometers, and LVDTs and was responsible for data acquisition. Financial support for the analysis of the test results was provided by the National Science Foundation under Grant No. CMS-0085353. This funding is gratefully acknowledged. The opinions and conclusions in this paper do not necessarily reflect the views of the sponsors.

REFERENCES

Ashford, S.A., Rollins, K.M., and Lane, J.D. (2004) "Blast-induced liquefaction for full-scale foundation testing," *J. Geotechnical and Geoenvironmetal Engrg.*, ASCE, Vol. 130, No. 8, 798-806.

Camp, W.M., Brown, D.A., and Mayne, P.W. (2002). "Construction method effects on axial drilled shaft performance" *Deep Foundations 2002, Geotechnical Special Publication No. 116*, ASCE, Vol. 1, p. 193-208.

Matlock, H., and Ripperger, E. A. (1958). "Measurement of soil pressure on a laterally loaded pile," *Procs. American Society for Testing Materials*. Vol. 58, 1245-1260.

Gerber, T. M. (2003). "P-y curves for liquefied sand subject to cyclic loading based on testing of full-scale deep foundations." Ph.D. Dissertation, Civ. & Environ. Engrg. Dept., Brigham Young Univ., Provo, Utah.

Hales, L.J. and Rollins, K.M. (2003). "Cyclic lateral load testing and analysis of a CISS pile in liquefied sand, GET Report 2003-5, Civ. & Environ. Engrg. Dept., Brigham Young Univ., Provo, Utah.

Kulhawy, F. H., and Mayne, P. W. (1990). Manual on estimating soil properties for foundation design. Research Project 1493-6, EL-6800, Electric Power Research Institute. Palo Alto, California.

Rollins, K.M., Gerber, T.M., Lane, J.D. and Ashford. S.A. (2005a). "Lateral resistance of a full-scale pile group in liquefied sand," *J. Geotechnical and Geoenvironmental Engrg.*, ASCE, Vol. 131, No. 1, p. 115-125.

Rollins, K.M., Lane, J.D., Dibb, E., Ashford, S.A., Mullins, A.G. (2005b). "Pore pressure measurement in blast-induced liquefaction experiments," Accepted for Publication, Transportation Research Record

Weaver, T.J., Ashford, S.A. and Rollins, K.M. (2005) "Lateral resistance of a 0.6 m drilled shaft in liquefied sand," *J. Geotechnical and Geoenvironmental Engrg.*, ASCE Vol. 131, No. 1, p. 94-102.

THE EFFECTS OF PILE FLEXIBILITY ON
PILE-LOADING IN LATERALLY SPREADING SLOPES

Stuart K. Haigh[1], S.P. Gopal Madabhushi[2]

ABSTRACT

Piles passing through laterally spreading slopes can be subjected to considerable loads by the soil flowing past them. Many case histories have been documented of piles which suffered failure as a result of horizontal loads exerted by the flowing soil.

This paper details the results of a series of dynamic centrifuge tests carried out at Cambridge University Engineering Department, to investigate the transfer of load from the spreading soil to the piles passing through it, with particular emphasis on the effective stress state of soil elements immediately upslope and downslope of the pile. This soil stress state can be calculated by virtue of instrumentation measuring both horizontal total stress and pore pressures at locations close to the upslope and downslope faces of the piles. By comparison of results obtained for both rigid and flexible piles, conclusions will be drawn as to the effects of pile flexibility on modifying the behavior of the soil-pile system.

INTRODUCTION

The many large earthquakes of recent years, including those at Northridge, Kobe, Taiwan and India, have highlighted the capacity of these events to cause massive destruction of infrastructure and huge loss of life. Whilst most fatalities in earthquakes are caused by structural failures, much of the damage to infrastructure comes from foundation failures, especially in areas where liquefaction is a possibility.

---

[1] Senior Engineer, Cambridge University, Cambridge, UK.
[2] Senior Lecturer, Cambridge University, Cambridge, UK.

Much of the expansion of the cities of the Pacific Rim over the past few decades has occurred by land reclamation from the sea, resulting in large areas of land that are extremely susceptible to liquefaction in areas of high seismic risk. The use of piles to carry structural loads through these strata to more competent ground is well established but if the ground is sloping and hence susceptible to lateral spreading following liquefaction, the interaction between these piles and the surrounding flowing liquefied soil is little understood. The capacity of these lateral forces to cause damage to pile-founded structures has been dramatically illustrated in many of the earthquakes of the past forty years.

Over the past decade, much research has been carried out investigating the interaction between piles and laterally spreading slopes with researchers such as Wilson (1998) and Brandenberg et al. (2004) all having studied the bending moments induced in piles by flowing soils, with and without non-liquefied crusts. This previous work tended to measure the bending moments induced in the piles, showing that maximum bending moment is found at the boundaries between liquefied and non-liquefied layers and then to back-calculate applied earth pressures. The work described here, in contrast, also directly measures the stresses applied to the upslope and downslope faces of the piles and hence allows a deeper insight into the soil behavior which regulates what forces are applied to the piles.

This paper describes the results of a series of dynamic centrifuge tests in which the interaction between flexible piles and laterally spreading soils was investigated by measuring both the soil's effective stress state close to the pile and the resulting bending moments induced in the pile. The results of these tests will be discussed and compared with those obtained from similar tests involving rigid piles. A full description of the results obtained from tests involving rigid piles is presented in Haigh and Madabhushi (2005).

The work described here concentrated on the effects of the lateral loading applied by the soil on the pile, and ignored the effects of pile axial loading on its performance. High axial load may have a significant effect in introducing instability of the pile due to p-δ effects. This aspect is dealt with in detail by Knappett & Madabhushi (2005)

CENTRIFUGE MODELLING

Centrifuge modeling is a technique by which the stress levels in a scale-model of a geotechnical structure can be made to be homologous to those in a full-scale prototype by testing the model within the enhanced g-field of a large centrifuge. As the model is at the correct stress level, non-linear constitutive effects are properly modeled and hence true prototype behavior is observed in the model.

The tests detailed here were carried out at the Schofield Centre of Cambridge University Engineering Department using the Turner 10m diammeter beam centrifuge. This centrifuge is described in detail by Schofield (1980) but briefly it consists of a 10m balanced beam with a radius to the model base of 4.125m, a payload of 1 ton and the capability to achieve 130g peak centripetal acceleration. In order to simulate earthquake shaking, the models described here were subjected to lateral shaking using a Stored Angular Momentum (SAM) actuator, described in detail by Madabhushi et al. (1998). This actuator uses the energy stored in spinning flywheels and released through a hydraulic clutch to impose unidirectional sinusoidal shaking to a model. The model in this case was enclosed within an Equivalent Shear Beam (ESB) model container, whose stiffness is chosen to match that of the enclosed soil layer in its non-liquefied condition, in order to minimize reflection of stress waves from the end walls of the model container during shaking. This model container is described in detail by Zeng and Schofield (1996), with the effects of the boundary conditions on liquefying soils within the container having been assessed by Teymur and Madabhushi, (2003).

A typical cross-section of the model discussed here can be seen in Figure 1. The model, a 1:50 scale replica of the prototype, which was tested at 50g, consists of a 6 degree slope of loose, (D_r=40%), liquefiable, fraction E silica sand, whose properties are summarized in Table A, overlying a dense (R_D=80%) base slope of the same sand. The loose sand layer is 100mm thick simulating a 5m thick layer in the prototype. At the top and bottom of the slope are permeable reservoirs of fraction B silica sand (whose properties are also shown in Table 1), in order to achieve plane-strain seepage through the model. The models are saturated with 50 cS silicone oil, in order to correct the anomaly between the scaling of dynamic and seepage time in centrifuge modeling, as discussed by Schofield (1981). It has been shown by Zeng et al. (1998) & Ellis et al (2000) that the use of silicone oil rather than water as a pore fluid has a negligible effect on the constitutive behavior of the soil for strain amplitudes of greater than 0.02%. Below this strain amplitude, some increase in damping is observed owing to the increased viscosity of the pore-fluid.

TABLE 1. Properties of Silica Sand

Property	Fraction B	Fraction E
ϕ_{crit}	36^0	32^0
D_{10}	0.84 mm	0.095 mm
D_{50}	0.9 mm	0.14 mm
D_{60}	1.07 mm	0.15 mm
e_{min}	0.495	0.613
e_{max}	0.82	1.014
G_s	2.65	2.65

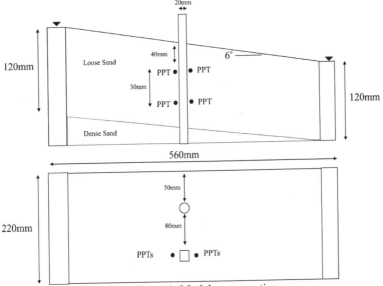

Figure 1: Model cross-section

In order to achieve true infinite slope behavior in a relatively short model, it was thought preferable to achieve slope-parallel seepage through the model slope. In order to achieve this, the water-table was maintained coincident with the slope surface at the top and bottom of the slope, with peristaltic pumps re-circulating the oil from the bottom of the slope to the top and maintaining steady state seepage. Owing to g-field curvature, it is impossible to maintain the water-table coincident with the surface along the entire length of the slope, but this error only results in the water-table dipping by 8 mm at the centre of the model.

The model contained two piles, of square and circular cross section, both having a diameter of 20mm (1m at prototype scale). These piles were constructed from a strain-gauged aluminum skeleton, surrounded by closed-cell foam. This enabled a very flexible pile to be constructed, having a prototype bending stiffness of 64MNm^2, while allowing both piles to have identical bending stiffness, the majority of this being contributed by the metal skeleton, rather than by the foam. The pile bases were inserted into clamps attached to the base of the model container in order to achieve a fixed-base boundary condition.

Instrumentation

The models were instrumented with D.J. Birchall Type A23 accelerometers, Druck PDCR81 pore-pressure transducers and Entran EPL contact stress cells. The instrumentation layout close to the pile is as shown in Figure 2. Whilst the circular pile was instrumented for bending moment with strain-gauge bridges, the near-pile pore-pressure and lateral stress gauges were concentrated around the square-section pile, as the flat faces of this pile allowed for better mounting of the stress cells and hence more accurate measurements of horizontal total stress to be achieved.

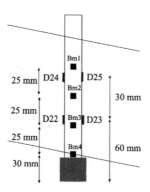

Figure 2: Pile Instrumentation

It can be seen from the figure that the instrumentation is concentrated close to the model pile. By measuring pore-pressure and horizontal total stress on both the upslope and downslope faces of the pile and by making assumptions about the other total stress components, it is possible to derive effective stress paths for soil elements close to the pile and hence to link the loading imposed on the pile with the soil constitutive behavior.

RESULTS

The centrifuge model was subjected to an approximately sinusoidal input motion, as shown in Figure 3. The fundamental frequency of this motion is 1 Hz, the duration 25s and the amplitude 0.2g. Whilst this input motion is a simplification of the true prototype scenario, having energy at only a limited number of frequencies, it is not thought that this would have a significant impact on the loading induced in the pile. The use of a simplified input motion for this experiment also allows the soil behavior to be much more easily extracted from the earth-pressure data.

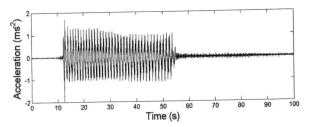

Figure 3: Input motion used for the experiments

Pore Pressures on Upslope and Downslope Faces of Piles

Pore pressures were measured approximately 10 mm from the upslope and downslope faces of the piles at the locations of the contact-stress cells. The pore pressures recorded at these locations are shown in Figure 4.

It can be seen that on both sides of the pile, very large "suction spikes" are observed due to suppressed dilation. This results in a variation of pore pressure of up to 30 kPa being observed in early cycles of the earthquake. This drop in pore pressure whilst flow is occurring obviously has a huge impact on the magnitude of the forces that can be applied to the piles by the flowing soil, as with 30 kPa of excess pore pressure the soil will act as a weak fluid, whereas with no excess pore-pressure, the soil is a relatively strong and stiff solid being forced past the pile. These suction spikes have been observed by other researchers such as Kutter & Wilson (1999) in laterally spreading soils, but the size of these spikes close to the pile is exceptional.

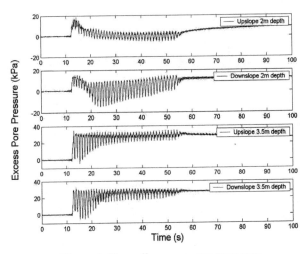

Figure 4: Near-pile excess pore pressures.

It can be seen that upslope of the pile at 2 m depth the pore pressure record shows that after the initial development of excess pore pressures sufficient to cause full liquefaction, rapid dissipation occurs from 15 to 20 s, after which cycling about the hydrostatic pore pressure is observed. This is possibly due to a high permeability drainage path forming between the pile and soil on the upslope side of the pile, allowing the excess pore pressures to dissipate close to the pile. Similar behavior is also observed downslope of the pile, though to a lesser extent. This may be due to the soil downslope of the pile being able to drain around the circumference of the pile and up through the high permeability path on the upslope side, more easily than draining directly to the surface.

The resultant downslope hydrodynamic pressure acting on the pile is shown in Figure 5. It can be seen that at 3.5 m depth, the pressure cycles about a positive value, whereas at 2 m depth the average value becomes more negative as the earthquake progresses. This is due to the drop in pore pressure observed through the earthquake upslope of the pile at 2 m depth.

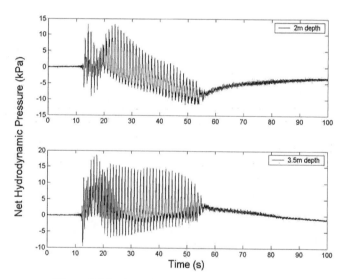

Figure 5: Net downslope hydrodynamic pressure

Lateral Earth Pressures on Upslope and Downslope Faces of Piles

The lateral earth pressures acting on the upslope and downslope faces of a square section flexible pile were measured using Entran EPL series stress cells. This pile had a prototype stiffness of 63.6 MNm^2. A typical 0.5m diameter steel tube pile would have a bending stiffness of about 150 MNm^2, so the pile modeled is slightly more flexible than might be the case in a field structure. Figure 6 shows time-histories of lateral pressure measured at depths of 2 m and 3.5 m on the upslope and downslope faces of the pile and Figure 7a shows the resultant downslope total stress acting on the pile at those depths. Figure 7b shows the comparable resultant downslope total stresses acting on very stiff solid brass piles subjected to the same earthquake motion. These piles had a prototype bending stiffness of 2000 MNm^2. Full details of the results of the tests involving rigid piles can be found in Haigh and Madabhushi (2005).

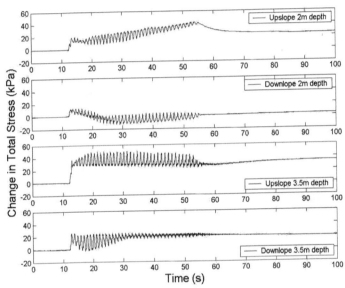

Figure 6: Horizontal total stresses acting on the flexible pile

It can be seen that a residual pressure difference of approximately 20 kPa is generated post-earthquake for both rigid and flexible piles, but that a peak transient value of approximately 50 kPa is generated during the earthquake itself for flexible piles. This peak transient value is approximately 40% higher at 70 kPa for rigid piles. It can also be seen that for flexible piles an increase in total horizontal earth pressure is observed during liquefaction on both faces of the pile, as would be expected owing to the earth-pressure coefficient approaching unity. The cyclic behavior, however, is largely contributed by large drops in earth-pressure on the downslope face with values approaching the initial values, whereas upslope of the pile increases in earth pressure from the post-earthquake value are observed. Looking at the timing of these transients reveals the reason for the complex and time-varying shape of the net downslope force measured at 2 m depth and shown in Figure 7. It can be seen from Figure 8 that early in the earthquake (from 20 s to 25 s) maximum upslope pressures and minimum downslope pressures occur simultaneously, whereas later in the earthquake (from 40 s to 45 s) maxima of both pressures begin to coincide.

It can also be seen that at locations downslope of the pile, early in the earthquake maximum horizontal total stresses and minimum pore-pressures occur simultaneously, whereas later in the earthquake pore-pressure and horizontal total stress changes are almost in-phase. This obviously significantly affects the shape of the stress path curves derived later in this paper.

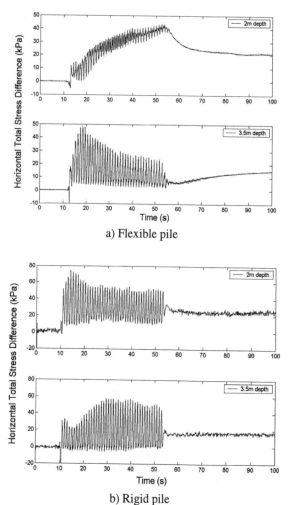

a) Flexible pile

b) Rigid pile

Figure 7: Net downslope total stress acting on flexible and rigid piles

Comparison of these results with those obtained from tests with rigid piles shows a much slower generation of the net horizontal earth pressure on the flexible pile. When the piles are rigid, the maximum horizontal earth pressure on the pile is achieved after about 3 cycles, whereas with the flexible pile, especially at low depth, there is a continuous increase in horizontal earth pressure with continued flow.

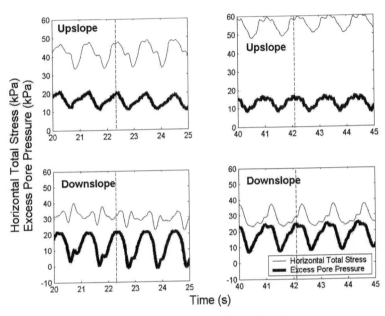

Figure 8: Relative phase of horizontal earth pressures

Stress Paths

Once horizontal total stress and pore-pressure are known, and assuming that the vertical total stress remains constant and equal to the overburden pressure at the instrument location, stress paths can be plotted upslope and downslope of the pile at the locations of the stress cells. These are shown in Figure 9.

It can be seen that upslope of the pile the stress path follows the passive failure line, whereas downslope of the pile it cycles between active and passive failure. This can be compared with the stress paths obtained from identical experiments on rigid piles, as seen in Figure 9b. It should be noted that at times the stress path is shown to cross the lines denoting active and passive failure. Active and passive failure are limits on static earth pressures, so it is possible that the earth pressures acting on the pile are transiently outside these bounds. It is also possible that some of the assumptions used to calculate these stress paths, such as vertical total stresses remaining constant, are violated hence distorting the stress paths predicted. The assumptions used were based on static equilibrium being maintained at all points during the earthquake, this is obviously not necessarily true during a dynamic event.

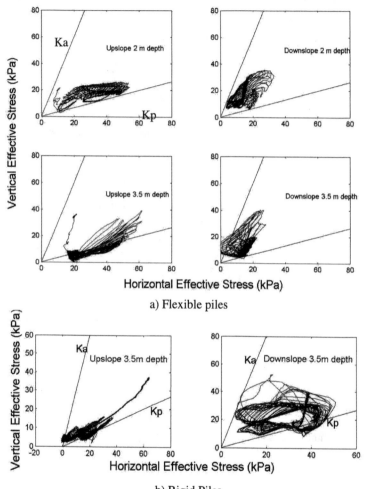

a) Flexible piles

b) Rigid Piles

Figure 9: Stress paths close to the square pile

It can be seen that with rigid piles, the greater relative soil-pile displacement causes the stress paths to cycle between active and passive pressure on both sides of the pile. This is caused by the pile flexibility, as the pile movement means that much more soil deflection is required before a steady state is reached. The accumulation of the stress difference is thus seen over a large number of cycles, rather than within a few cycles, as was seen from the stress cycles for the rigid piles.

Bending Moments

The bending moments recorded by the strain gauges within the square and circular piles are shown in Figure 10. It can be seen that whilst the residual bending moments measured in the circular and square piles are comparable, those in the square pile being larger by about 8%, the rate of generation of bending moment is significantly greater in the square pile. After 10 s of the earthquake, bending moments in the square pile exceed those in the circular pile by 30%. This implies that less soil displacement is required to mobilize the full horizontal loading on the square piles than on the circular piles. Thus, if only limited displacement of the soil had occurred it might be expected that the bending moments in the square pile would significantly exceed those in the circular pile.

It is also obvious that the bending moments measured in the square pile at 1.25 m depth show much greater cyclic behavior than those in the circular pile.

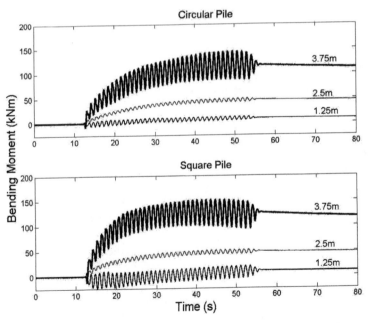

Figure 10: Bending moments recorded in circular and square piles

A comparison between the measured bending moments and applied lateral stresses can be found in the next section. It will be seen that a pseudo-static application of the applied loads would lead to greater bending moments being sustained by the pile than the are measured due to their dynamic application.

IMPLICATIONS TO DESIGN CODES

Comparison of both the horizontal earth pressures exerted on the piles and the bending moments induced in the piles can be made with those suggested by the JRA highway bridges design code (JRA, 1996). It can be shown that whilst the bending moments exerted in the flexible piles exceed by a factor of 2 those suggested by the design codes, the peak lateral loadings on the piles can exceed the suggested values by up to a factor of five. A comparison of the bending moments and lateral loads measured in these experiments and predicted by the JRA code is shown in Figure 11.

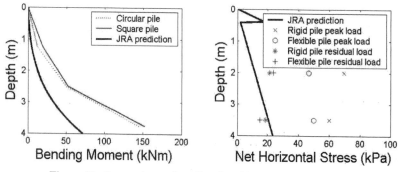

Figure 11: Comparison of predicted and measured pile loading

Whilst this might seem contradictory, it must be remembered that the JRA design code is for pseudo-static design and hence if earth pressures at different depths become out of phase, this dynamic factor will result in the peak applied loadings at all depths not occurring simultaneously and hence a lower bending moment being suffered than would be the case if all earth pressures varied in phase.

CONCLUSIONS

It has been shown that dynamic centrifuge modelling is a powerful tool for the study of the behaviour of pile foundations under seismic loading. Comparison of the results obtained from the study of rigid and flexible piles within laterally spreading slopes has shown the greater forces imparted on rigid piles than on flexible piles by the flowing ground. It has also been shown that whilst the JRA design code predicts sensible values of applied bending moment for very flexible piles, when very rigid piles are used, the design values may be a significant underestimate of the applied loading.

REFERENCES

Brandenberg, S.J., Boulanger, R.W., Kutter, B.L., Wilson, D.W. and Chang, D. (2004). "Load transfer between pile groups and laterally spreading ground during earthquakes", Proc. 13th World Conference on Earthquake Engineering, Vancouver, Canada, Paper No. 1516

Ellis, E.A., Soga, K., Bransby, M.F. and Sato, M. (2000). "Resonant column testing of sands with different viscosity pore fluids", ASCE Journal of Geotechnical and Geoenvironmental Engineering, Vol.126, No.1, pp.10-17

Haigh and Madabhushi (2005). "Pile-soil interaction in liquefiable slopes", Submitted to ASCE Journal of the Geotechnical and Geoenvironmental division (under review)

Japan Road Association (1996). "Specifications for highway bridges, part V: Seismic design", pp. 90-95

Knappett, J.A., and Madabhushi, S.P.G. (2005). "Modelling of liquefaction induced instability in pile groups", Proc. Int. Workshop Simulation and seismic performance of pile foundations in liquefied and laterally spreading ground, ASCE.

Kutter, B.L., and Wilson, D.W. (1999). "De-liquefaction shock waves", Proc. 7th US-Japan Workshop on Earthquake-Resistant Design of Lifeline Facilities and Countermeasures Against Soil Liquefaction, Seattle, WA, pp. 295-309.

Madabhushi, S.P.G., Schofield, A.N. & Lesley, S. (1998). "A new stored angular momentum based earthquake actuator", Proc. Centrifuge '98, Tokyo, Vol.1, pp. 111-116

Schofield, A.N. (1980). "Cambridge geotechnical centrifuge operations", Géotechnique, Vol.25, No.4, pp. 743-761

Schofield, A.N. (1981). "Dynamic and earthquake geotechnical centrifuge modelling", Proc. Int. Conf. on Recent Advances in Geotechnical Earthquake Engineering and Soil Dynamics, St Louis, Vol.III, pp. 1081-1100

Teymur, B., and Madabhushi, S.P.G., (2003). "Experimental study of boundary effects in dynamic centrifuge modelling", Géotechnique, Vol.53, No.7, pp.655-663

Wilson, D.W. (1998) "Soil-pile-superstructure interaction in liquefying sand and soft clay", PhD Thesis, University of California at Davis, CA

Zeng, X. and Schofield, A.N. (1996). "Design and performance of an equivalent shear beam container for earthquake centrifuge modelling", Géotechnique, Vol.46, No.1, pp.83-102

Zeng, X., Wu, J. and Young, B.A. (1998). "Influence of viscous fluids on properties of sand", ASTM Geotechnical Testing Journal, Vol.21, No.1, pp.45-51

SOIL-PILE INTERACTION IN HORIZONTAL PLANE

Susumu Iai[1], Tetsuo Tobita, Member, ASCE[1], Matthew Donahue[2], Masato Nakamichi[3], and Hidehisa Kaneko[3]

ABSTRACT

Two dimensional model tests are performed on a horizontal cross section of a soil-pile system in a pile foundation. The objective of the model tests is to evaluate local soil displacement field in the vicinity of the piles associated with a global displacement of soil around the pile foundation. Two dimensional effective stress analyses in horizontal plane are also performed to generalize the findings from the model tests. An effective stress model based on multiple shear mechanism is used through a computer code FLIP. Primary findings from this study are as follows:

(1) In dry condition, displacement vectors are directed away from pile front, and displacement at pile side rapidly decreases with an increasing distance from soil-pile interface. In undrained condition, displacement field shows vortexes at pile side associated with push-out/pull-in pattern of displacements in front of and behind the pile.

(2) Distribution of local soil displacement between piles deployed perpendicular to direction of global displacement of soil shows high strain concentration (i.e. discontinuity in displacement) at soil-pile interface.

INTRODUCTION

Soil-pile interaction associated with global movement of soil such as shown in Figure 1 is associated with a highly nonlinear phenomenon. In the case history

---

[1] Disaster Prevention Research Institute, Kyoto University, Kyoto, Japan.
[2] Oregon State University, U.S.A.
[3] Kinki Regional Development Bureau, Ministry of Land, Infrastructure and Transport, Japan

shown in this figure, slope under a deck of a pile-supported wharf moved 1.5m towards the sea, causing buckling at deck-pile connection and at embedded portion of piles (Iai, 1998). Although soil-pile interaction with small displacement/strain may be relatively easily analyzed using three dimensional finite element technique, soil-pile interaction with large displacement/strain may pose a challenge to engineers and researchers. For example, highly non-linear nature of soil can result in strain concentration at soil-pile interface, posing difficulty in numerical analysis that may have worked well in linear analysis. At large displacement/strain level, much finer finite element mesh may be needed in the vicinity of soil-pile interface. Numerical robustness in numerical analysis may be needed for simulating the highly non-linear soil-pile interaction. Although three dimensional finite element technique incorporating non-linear behavior of soil has been started to be used for numerical simulation of soil-pile interaction problems (e.g. Kimura et al, 2000; Sato et al, 2003), how soil behaves in the vicinity of piles still remains the issue to be studied.

In this study, two dimensional model tests are performed on a horizontal cross section of a soil-pile system. The objective of the model tests is to evaluate a local soil displacement field in the vicinity of piles such as that illustrated in Figure 2. Two dimensional effective stress analyses in horizontal plane are also performed to generalize the primary findings from the model tests.

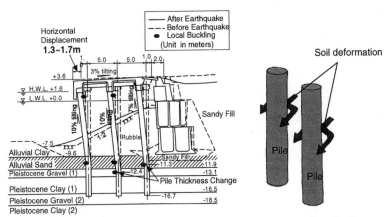

Figure 1. Damage to a pile-supported wharf during Hyogoken-Nambu earthquake of 1995 (after Iai, 1998)

Figure 2. Schematic figure of soil deformation around piles

MODEL TESTS FOR SOIL DEFORMATION AROUND PILES

The model tests were performed using an aluminum container (inner dimensions: 800mm long, 500mm wide, 40mm high), in which a cylindrical pile model made of Teflon, 40mm high with a diameter of 50mm, was embedded in a sand deposit formed in the container as shown in Figure 3. The sand deposit was formed by air pluviation for dry condition, and by poring a slurry mixture of sand and viscous fluid (120cSt) for saturated condition. Silica No.7 sand was used. Relative densities of the sand deposits were about 70% for dry condition and about -150% (negative relative density) for saturated condition. After the sand deposit was formed, an acrylic plate was placed on the surface on the sand deposit. Displacement was induced to the pile model by pulling a wire attached to the mid portion of the pile model at the rate of 7.2mm/min. Although the pile model was moved in the model tests, the primary interest of the model tests was to measure the displacement field of soil relative to the movement of the pile. Thus, the results of the model tests are readily applicable to the conditions when the global soil movement is induced around the pile foundation as shown in Figure 1.

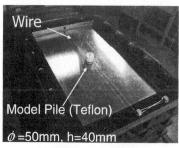
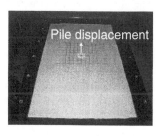

(a) Before poring sand   (b) After forming sand deposit
Figure 3. Apparatus for model tests for soil-pile interaction in horizontal plane

Eight cases of model tests were performed with a varying combination of pile conditions (single or group pile) and soil conditions (dry or saturated). Model tests using a sand paper attached around the model pile were also performed for studying the effect of skin friction at soil-pile interface (Cases-2 and 4). Group pile tests were performed using a combined model of piles shown in Figure 4, where the pile models

Table 1. Conditions for model tests

Case No.	Pile	Sand deposit	Relative density (%)
Case-1	Single	Dry	70
Case-2	Single (friction large)		
Case-3	Single	Saturated	-135
Case-4	Single (friction large)		
Case-5	Group (3D)	Dry	72
Case-6	Group (3D)	Saturated	-166
Case-7	Group (5D)	Dry	66
Case-8	Group (5D)	Saturated	-161

Figure 4. Pile model for group pile

were connected with a brass rod, 4mm in diameter, along the direction of displacement with pile spacings of 3D and 5D.

The local displacement field in the vicinity of the pile was monitored through the acrylic plate using a video-camera and a digital microscope. Load applied on the pile model was measured using a load cell attached at the end of the wire outside the container. Displacement of the pile model was measured using a laser transducer for monitoring the displacement at the end of the wire outside the container.

MEASURED DISPLACEMENT FIELD OF SOIL AROUND SINGLE PILE

The local displacement field monitored through a video-camera was plotted in terms of displacement vectors at nodes of the grid formed by colored sand markers. Under dry condition (Case-1), the displacement vectors were directed away from the front of the pile in a pattern of a fan as shown in Figure 5. The displacement vectors at pile side rapidly decreased with an increasing distance from soil-pile interface. A void was formed behind the pile following the movement of the pile. Under saturated condition (Case-3), vortexes were formed at pile side as shown in Figure 6. Void formation was not observed behind the pile under saturated condition.

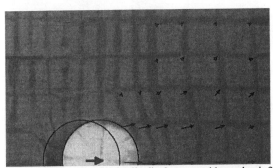

Figure 5. Measured displacement field (pile displacement 11mm, load=20N)
(Case-1: dry)

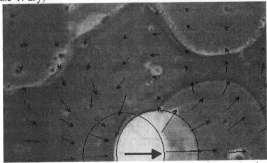

Figure 6. Measured displacement field (pile displacement 21mm, load=6N)
(Case-3: saturated)

A zoomed up displacement field of soil at the soil-pile interface was measured using a digital microscope as shown in Figure 7. Individual sand particle was used as a target to produce the displacement vector. From this measurement, displacement distribution in the vicinity of soil-pile interface was obtained as shown in Figure 8.

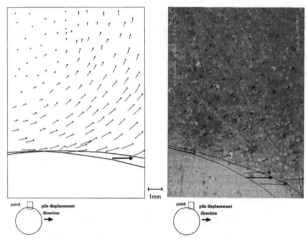

Figure 7. Measured displacement field in the vicinity of a pile in dry condition (Case-1)

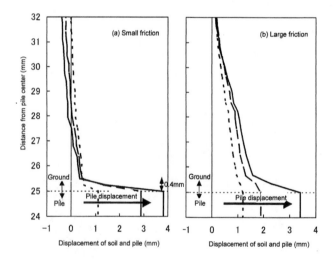

Figure 8. Displacement distributions in the vicinity of soil-pile interface for dry sand deposit (Cases-1 and 2)

As shown in Figure 8 (a), variation in displacement is concentrated at the soil-pile interface when skin friction of the pile was small (Case-1). When skin friction of the pile becomes large (Case-2), displacement of soil was affected over larger distance from the soil-pile interface as shown in Figure 8 (b).

Similar results are obtained for saturated sand deposit as shown in Figure 9. Under saturated condition, the effect of friction was more clearly recognized than in dry sand deposit.

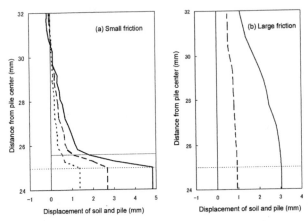

Figure 9. Displacement distributions in the vicinity of soil-pile interface for saturated sand deposit (Cases-3 and 4)

MEASURED DISPLACEMENT FIELD OF SOIL AROUND PILE GROUP

The measurements similar to those for single pile were also made for a pile group. As shown in Figure 10, the displacement field in front of the front pile of the pile group at dry sand deposit (Case-5) was basically the same as that for single pile (Case-1) shown in Figure 5. The soil between the front and middle piles in the pile group showed uniform displacement field, moving as a body united with the group. Displacement field at the side (upper or lower part of soil in the figure) of the pile group was less affected by the movement of the pile group.

For saturated sand deposit (Case-6), displacement vectors showed a pattern of a fan between the front and following piles and differ from those at dry sand deposit. At the side of the pile, vortexes were recognized that may be associated with the displacement of soil that was pushed out in front of and pulled in behind each pile.

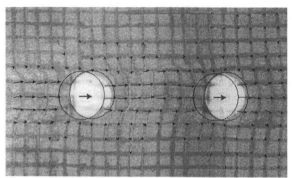

Figure 10. Displacement vectors at dry sand deposit (Case-5) (between the front and the middle pile) (pile displacement 15mm, load 90N)

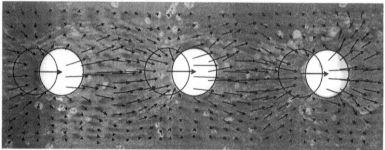

Figure 11. Displacement vectors at saturated sand deposit (Case-6) (pile displacement 30mm, load 3N)

ANALYSIS OF A SOIL-PILE SYSTEM IN HORIZONTAL PLANE

Two dimensional analysis of a horizontal cross section of the soil-pile system was performed under pseudo-static conditions. An effective stress model based on multiple shear mechanism was used through a computer code FLIP (Iai et al, 1992). In this analysis, a single row of equally spaced piles deployed perpendicular to the direction of load (Figure 12(a)) was idealized into an analysis domain defined by the boundaries that run parallel to the load direction and go through the centers of the pile spacing. These boundaries were periodic, sharing the same displacements at the boundary nodes with the same x-coordinate, where x-axis is directed towards right on the paper. At the right and left side boundaries on the paper, x-displacements were fixed.

Finite element mesh used for the analysis of a single row of piles with a spacing of $L=10D$ and a pile diameter $D=5$cm is shown in Figure 13 for the area ranging from $L=-5D$ to $+5D$. In the analysis, whole soil-pile system was initially consolidated with a confining pressure of 0.28 kPa for simulating the confining condition at the middle depth of the model sand deposit (i.e. 2cm from the surface). The cylindrical pile section was idealized using linear solid elements. This pile section was replaced by the soil elements in the initial phase of analysis for consolidation in order to avoid artificial stress concentration. Following this initial phase, the pile was loaded with a monotonically increasing load. Soil deformation around the cylindrical cross section of the pile was computed in drained and undrained conditions. Parameters for sand used for the analysis were determined referring to the results of laboratory tests on Silica sand No.7 as shown in Table 2.

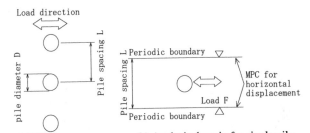

(a) Pile rows   (b) Analysis domain for single pile
Figure 12. Two dimensional analysis of a soil-pile system in horizontal plane

Figure 13. Finite element mesh used for the analysis

Table 2. Parameters for silica sand No.7

ρ_t (t/m³)	G_{ma} (kPa)	ν	σ_{ma}' (kPa)	ϕ_f (deg)	H_{max}
2.0	3760	0.33	0.28	35	0.240

ρ_t: density; G_{ma}: initial shear modulus at a confining pressure of σ_{ma}'; σ_{ma}': reference confining pressure; ϕ_f: internal friction angle; ν : Poisson's ratio; H_{max}; limiting value of hysteretic damping factor (ϕ_p :phase transformation angle and w_1, p_1, p_2, c_1, s_1: parameters for dilatancy were not used.)

COMPUTED RESULTS FOR SINGLE PILE

The numerical analysis was performed in a stable manner even in the order-of-magnitude displacement comparable to that to the pile diameter. Numerical instability occurred only when the ultimate state of the soil was reached either when the load resistance curve of the pile reached a plateau, or when there was instability in the soil under drained condition.

Computed displacement field for the dry condition is shown in Figure 14. The displacement vectors are directed away from the front of the pile in a pattern of a fan. The displacement vectors at the pile side rapidly decreases with an increasing distance from the soil-pile interface. In order to clearly show the displacement distribution between the piles, horizontal components of the displacements are plotted in Figure 15. These results are basically consistent with those measured and shown in Figures 5 and 8. Only difference is noted with respect to the formation of voids behind the pile in the model tests. No void was formed in the analysis probably because the confining stress in the analysis was more uniform than the one in the model tests where the stress field became 3-D when the void began to form.

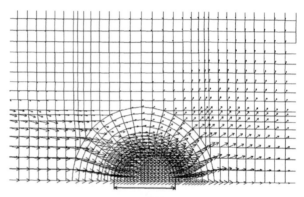

Figure 14. Computed displacement field around pile (drained) (Case-2)

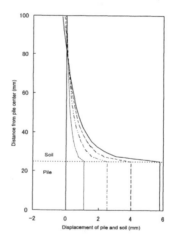

Figure 15. Computed displacement distributions between the piles (drained) (Case-2)

Computed displacement field for the undrained condition is shown in Figure 16. Displacement vectors beside the pile shows vortexes. In order to clearly show the displacement distribution between the piles, x-components of the displacements are plotted in Figure 17. These computed results are basically consistent with those measured and shown in Figures 6 and 9. Only difference noted is with respect to the manner in which the displacements are decaying from the center of the pile: the model tests with large friction shows much slower rate of decay at the soil-pile interface than the analysis. A further study may be needed to follow up this issue.

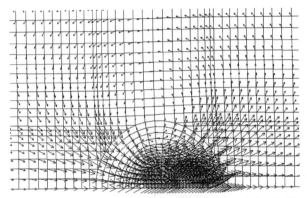

Figure 16. Computed displacement field around pile (undrained) (Case-4)

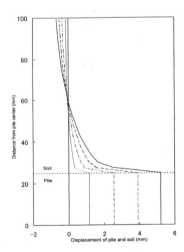

Figure 17. Computed displacement distributions between the piles (undrained) (Case-4)

COMPUTED RESULTS FOR PILE GROUP

Analysis of the pile group of three rows was also performed. In the analysis, the three rows of piles underwent the same displacement in horizontal direction. As an example of analysis results, deformation and displacement vectors around the pile group computed for undrained conditions are shown in Figure 18. The computed results are consistent with those measured and shown in Figure 11.

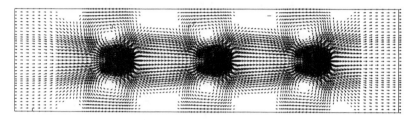

Figure 18. Computed displacement vectors around the pile group for undrained condition (Case-6)

CONCLUSIONS

Two dimensional model tests and effective stress analysis were performed on a horizontal cross section of a soil-pile system. Major findings from this study are as follows:

(1) Local displacement field of soil in the vicinity of a pile associated with global displacement of soil around pile foundation shows marked difference depending on dry/saturated conditions. In dry condition, displacement vectors are directed away from the pile front, and displacement at the pile side rapidly decreases with an increasing distance from the soil-pile interface. In undrained condition, displacement field shows vortexes at the pile side associated with the push-out/pull-in displacements in front of and behind the pile.

(2) Local displacement field of soil for group pile in dry condition shows that the soil between the front and following piles of group pile moves with the piles whereas the soil beside the pile group moves independently from the movement of the soil-group pile mass.

(3) Local displacement field of soil for group pile in undrained condition shows a pattern of a fan-out/fan-in between the front and following piles and differs from those at drained condition. At the side of the pile group, vortexes are clearly recognized that may be associated with the displacement of pushed-out/pull-in pattern of displacement field of soil at each pile.

(4) Distribution of soil displacement between the piles deployed perpendicular to the direction of global displacement of soil shows high strain concentration (i.e. discontinuity in displacement) at soil-pile interface. Distribution of soil displacement around the soil-pile interface is strongly affected by skin friction of

the pile.

(5) The computer code FLIP for effective stress analysis shows numerical robustness in nonlinear soil-pile interaction analysis: no difficulty is encountered in the numerical analysis of the soil-pile system involving large displacements in the order-of-magnitude displacement comparable to that of the pile diameter.

(6) The computed displacement fields are basically consistent with those measured, supporting the major findings summarized in (1) through (4) obtained from the model tests.

REFERENCES

Iai, S., Matsunaga, Y., and Kameoka, T. (1992). "Strain space plasticity model for cyclic mobility." *Soils and Foundations*, Vol.32, No.2, pp.1-15

Iai, S. (1998). "Seismic analysis and performance of retaining structures." *Geotechnical Earthquake Engineering and Soil Dynamics III, Geotechnical Special Publication No.75, ASCE*, Vol.2, pp.1020-1044

Kimura, M. and Zhang, F. (2000). "Seismic evaluations of pile foundations with three different methods based on three-dimensional elasto-plastic element analysis." *Soils and Foundations*, Vol.40, No.5, pp.1-17

Sato, T., Matsumura, T., Zhang, F., Moon, Y., and Uzuoka, R. (2003). 3-dimensional simulation of pile-group system during liquefaction and following ground flow process." *Proc. 8ᵗʰ U.S.-Japan Workshop on Earthquake Resistant Design of Lifeline Facilities and Countermeasures against Liquefaction*, Technical Report MCEER-03-0003, pp.507-518

EFFECT OF SOIL PERMEABILITY ON CENTRIFUGE MODELING OF PILE RESPONSE TO LATERAL SPREADING

Lenart González, S. M. ASCE[1], Tarek Abdoun, A. M. ASCE[1], Ricardo Dobry, M. ASCE[1]

ABSTRACT

Liquefaction-induced lateral spreading continues to be a major cause of damage to deep foundations. Currently there is a huge uncertainty associated with the maximum lateral pressures and forces applied by the liquefied soil to deep foundations. Furthermore, recent centrifuge and 1g shaking table tests of pile foundations indicate that the permeability of the liquefied sand is an extremely important and poorly understood factor. This article presents experimental results and analysis of one of the centrifuge tests that were conducted at the 150 g-ton RPI centrifuge to investigate the effect of soil permeability in the response of single piles and pile groups to lateral spreading.

INTRODUCTION

Liquefaction-induced lateral spreading of sloping ground and near waterfronts continues to be a major cause of damage to deep foundations. In the US, Japan and other countries, buildings, bridges, and other structures supported by deep foundations have been damaged in many earthquakes, with billions of dollars in damages. Permanent lateral ground deformations induce cracking and rupture of piles at both shallow and deep elevations, rupture of pile connections, and permanent lateral and vertical movements and rotations of pile heads with corresponding effects on the superstructure (McCulloch and Bonilla, 1970; Hamada et al., 1986; Mizuno, 1987; Hamada and O'Rourke, 1992; O'Rourke and Hamada, 1992; Youd, 1993;

---

[1] Rensselaer Polytechnic Institute, Troy NY, USA

50

Swan et al., 1996; Ishihara et al., 1996; Tokimatsu et al., 1996; Yokoyama et al., 1997; Tokimatsu, 1999; Dobry and Abdoun, 2001).

While in some cases the top of the foundation displaces laterally a distance similar to that in the free field, in others it moves much less due to the constraining effect of the superstructure, or of the deep foundation's lateral stiffness including pile groups and batter piles. The foundation may be exposed to large lateral soil pressures, including especially passive pressures from the nonliquefied shallow soil layer riding on top of the liquefied soil. In some cases, this soil has failed before the foundation with negligible bending distress and very small deformation of the foundation head and superstructure (Berrill et al., 1997); while in others the foundation has failed first in bending and/or has experienced excessive permanent deformation and rotation at the pile heads. The observed damage and cracking to piles is often concentrated at the upper and lower boundaries of the liquefied soil layer where there is a sudden change in soil properties, or at the connection with the pile cap. More damage tends to occur to piles when the lateral movement is forced by a strong nonliquefied shallow soil layer, than when the foundation is free to move laterally and the forces acting on them are limited by the strength of the liquefied soil.

CURRENT PRACTICE AND UNCERTANTIES

Case histories, as well as 1g shaking table and centrifuge model tests, indicate that the effect of lateral spreading on piles can be characterized in first approximation as a pseudostatic, kinematic soil-structure interaction phenomenon, driven by the permanent lateral movement of the ground in the free field. Various foundation analysis and design methods have been proposed, where the soil applies static lateral forces to the pile foundation, either (i) as a function of the relative displacement between the foundation and the free field (p-y approach); or (ii) taking the maximum possible values of these lateral soil static forces which depend on the soil strength within an overall limit equilibrium (LE) method. A third approach (iii) suggested by several Japanese researchers assumes that the liquefied soil is a viscous fluid and hence the lateral soil static forces are a function of the relative velocity, rather than the relative displacement, between foundation and free field (Hamada, 1998; Higuchi and Matsuda, 2002).

There is currently a huge uncertainty associated with the maximum lateral pressures and forces applied by the liquefied soil, which translates into a similar huge uncertainty in the calculated maximum pile bending moments. For example, in the Japan Road Association (JRA) method, the lateral pressure is specified as 30% of the total overburden pressure, while Abdoun et al. (2003) has recommended a constant lateral pressure with depth of 10 kPa. For a range of field conditions involving single piles (but not necessarily pile groups), the JRA and Abdoun method give similar results. A main source of uncertainty is the area over which this pressure is applied in the case of pile groups. Yokoyama et al. (1997) suggests that the value of the lateral pressure must be multiplied for the whole area of the pile group including the soil between the piles, which for a pile separation of 3d (d = pile diameter) may give a lateral force as much as three times greater than if the lateral pressure is applied only to the piles.

Furthermore, recent centrifuge and 1g shaking table tests (small and full scale) of single piles and pile groups indicate that the permeability of the liquefied sand is an extremely important and poorly understood factor. Several researchers have found out recently that the resistance of the liquefied soil to the movement of an object (pile, cylinder, or sphere) increases as the relative velocity of the object and the soil increases. These results support the theory that the liquefied soil can be modeled as a viscous fluid (e.g. Dungca et al., 2004; De Alba and Ballestero, 2004; Hwang et al., 2004). Dungca et al. conducted small-scaled shaking table tests to study the lateral resistance of a pile subjected to liquefaction-induced lateral flow, where he modeled the pile as a buried cylinder. The results support that the pore fluid migration rate, i.e. the hydraulic conductivity of the soil with respect to the loading rate, is the crucial factor for mobilization of the lateral resistance of a buried cylinder in liquefied soil, because there is less time for the pore fluid to come rushing from the free field to dissipate the negative pore pressures near the pile or other object.

A series of centrifuge tests were conducted the 150 g-ton RPI centrifuge to investigate the effect of soil permeability in the response of single piles and pile groups to lateral spreading. More specifically, six models, simulating a mild infinite slope with a liquefiable layer on top of a nonliquefiable layer were tested in a large laminar box. One model consisted of a single pile (model 1x1-w), other consisted of a line of three piles with a pile cap perpendicular to the direction of the lateral spreading (model 3x1-w) and a third model consisted of a pile group of 2x2 with a pile cap (model 2x2-w). These three models were tested using water as the pore fluid. All models were repeated, using the same fine sand, but saturated this time with viscous fluid (models 1x1-v, 3x1-v and 2x2-v respectively), hence simulating two sands of widely different permeabilities in the field. The importance of the permeability of the liquefied sand can be illustrated comparing the models that simulated the 2x2 pile group. In the model 2x2-w (saturated with water), the pile cap reached a maximum lateral displacement of 7 cm and a maximum bending moment of 55 kN-m in prototype units and then bounced back, while in the model 2x2-v (saturated with viscous fluid) the pile cap reached a maximum displacement of 45 cm and a maximum bending moment of 425 kN-m at the end of shaking, without ever bouncing back. This is a factor of 375/55 $\approx$ 7 between maximum pile bending moments. Therefore, the uncertainty in lateral soil forces and pile bending moments, related to the poor understanding of the complex behavior of liquefied soils in the vicinity of foundations, can produce maximum lateral liquefied soil forces and pile bending moments varying by factors as high as 3 or 7. It is necessary to reduce this huge uncertainty to more reasonable values in order to develop rational methods of analysis and design of deep foundation subjected to lateral spreads.

CENTRIFUGE MODELING OF SINGLE PILE RESPONSE TO LATERAL SPREADS

This paper presents results and analyses of the centrifuge test corresponding to the model 1x1-v of the experimental study mentioned above. Figure 1 presents a sketch of RPI's large laminar box, soil profile, single pile, and instrumentation used in this model. The model height was approximately 0.16 m, simulating under a 50-g level an 8 m prototype soil deposit. The profile consisted of a 6 m layer of loose Nevada sand placed at a relative density of about 40 %, overlying a 2 m layer of slightly cemented sand. The pile was embedded into the nonliquefiable layer, simulating a end-bearing pile. The nonliquefiable layer, which consisted of the same Nevada sand but slightly cemented, was placed by dry pluviation and then saturated with water. After 24 hours the liquefiable layer was also placed by dry pluviation and then saturated with a water-metulose solution with a viscosity of about 50 times the viscosity of water, hence simulating a fine sand deposit in the field.

The pile was placed in the model before the soil was pluviated, attempting to simulate a pile installed with minimal disturbance to the surrounding soil, as may be the case when a pile is inserted into a pre-augered hole. The pile consisted in a 0.95 cm diameter polyetherimide rod, simulated at 50-g a prototype pile diameter of 47.5 cm with a bending stiffness (EI) of 9000 kN-m^2. After placing the strain gauges at different locations along the pile to measure bending moments, the pile was covered with a thin layer of wax and a soft shrink tube. Then, sand grains were glued to the side of the pile to develop an adequate pile-soil roughness representing the interface between soil and a reinforced concrete pile. The final effective prototype pile diameter was approximately 60 cm. Besides the strain gauges to measure bending moments, LVDTs were installed at the top to measure the pile head displacement. The soil was instrumented with pore pressure transducers and accelerometers, as well as with lateral LVDTs mounted on the rings of the flexible wall to measure soil deformations in the free field. Grids of colored sand were placed at intermediate

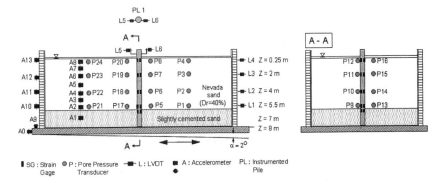

Figure 1: Sketch of laminar box and instrumentation used in models 1x1-w & 1x1-v

depths to observe the pattern of soil displacement around the pile at the end of shaking. The model was inclined 2° to the horizontal, which simulated an inclination of 4.8° after pertinent corrections, thus simulating a mild infinite slope. The models were excited in flight with an input base acceleration (Fig. 2) consisting of 30 cycles of uniform acceleration having a prototype amplitude of around 0.3 g and a frequency of 2 Hz. Experimental results and detailed interpretation of the model are presented in the following section.

EXPERIMENTAL RESUTLS AND ANALYSIS

The input acceleration (recorded) and recorded accelerations in the soil at different depths during shaking are shown in Fig. 2. The corresponding accelerometers were located at reasonable distances from the piles, so these accelerations can be considered as free field data. The acceleration records at the ground surface and at a depth of 2 m decreased significantly after about 1 cycle of shaking due to the liquefaction process and the dynamic isolation of the shallower layers. The acceleration records contain large spikes in each cycle due to the dilative behavior of the saturated loose layer during lateral spreading. The slightly cemented sand acted as a solid layer during shaking, as illustrated by the acceleration records being essentially identically to the input acceleration.

A large number of pore pressure transducers were placed in the model, far away, close and next to the pile, as shown in Fig. 2. These measurements are very important to understand the effect of fluid viscosity in the response of the pile foundations. In the free field, the excess pore pressure records (P1, P2, P3, P4) reveal that the soil liquefied after about one or two cycles of shaking (Fig. 3), in agreement with the trend exhibited by the acceleration time histories. However, near the ground surface the excess pore pressure decreased after a couple of cycles of shaking to values close to cero. Large shear strains developed under low confinement and a slow dissipation process appear to be responsible for this phenomenon. Figure 3 also shows the excess pore pressure measured next to the pile. Negative excess pore pressure developed also near the ground surface (P12); however, in this case the tendency was much stronger, reaching values up to -20 kPa at the end of shaking. The decrease in lateral stress on the downslope side of the pile, and large shear strains with an undrained dilative response of the liquefied soil close to the pile seam to have been responsible for this phenomenon. At larger depths the pore pressure records do not exhibit such a dramatic response. The development of negative excess pore pressure next to the pile generated a large vertical hydraulic gradient, which caused a faster pore pressure dissipation than a greater distance from the pile or in the free field.

Profiles of the free field lateral displacement were obtained interpolating the values measured with the LVDTs placed in the laminar rings (Fig. 4). As soon as the loose sand liquefied at the beginning of shaking, the deposit started moving laterally downstream, reaching a maximum displacement at the end of the excitation of 140 cm. The pile displacement profiles, without the dynamic component, were obtained

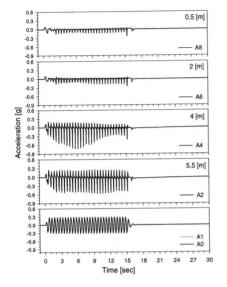

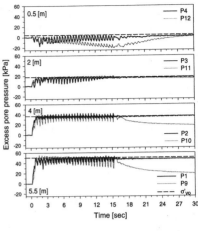

Figure 2: Acceleration time histories

Figure 3: Excess pore pressure time histories

as a first approximation through double integration of the interpolated profiles of bending moments along the height, according to the equation:

$$y_p = \iint \frac{M(h)}{EI} \, dh \qquad [1]$$

where y_p is the pile lateral deformation, EI is the flexural rigidity of the pile, and h is the distance from the slightly cemented layer (height). However, the estimated pile head lateral displacement differed considerably from the measured with the LVDTs L5 and L6, due to the fact that the slightly cemented was not able to provide infinite lateral constraint. In a second iteration, the angular rigidity was estimated dividing the bending moment measured at the base of the liquefiable layer by the necessary rotation, so the pile head displacement obtained with equation [1] plus the displacement due to the rotation match the displacement measured with the LVDTs. It was obtained that a value of 8000 kN-m/rad represents very good the angular rigidity of the slightly cemented layer (González et al., 2005). Finally, considering the deformation by curvature and rotation, the profiles of lateral displacement were obtained (Fig. 4), which increase monotonically during shaking, reaching a maximum displacement of 65 cm at the end of shaking.

As already mentioned, grids of colored sand were placed at intermediate depths to observe the effect of soil permeability in the pattern of soil displacement around the pile. Figure 5 shows a picture of soil condition at 1 m depth, taken after the test. The arrow indicates the direction of lateral spreading. The picture shows a large area of influence due to the presence of the pile, reaching a distance of several times the diameter of the pile.

Bending moment profiles measured at different times in the single pile, after filtering out the dynamic component, are shown in Fig. 6. The pile reached a maximum bending moment of 360 kN-m at the base of the liquefiable layer at the end of the excitation, which should be compared to the model 1x1-w saturated with water (not presented in this article), where the pile bounced back during shaking reaching a maximum bending moment much smaller (120 kN-m).

In order to further investigate the liquefied soil-pile-structure interaction during lateral spreading, the lateral soil resistance (p) was estimated using the simple shear beam theory, according to equation [2], where z is the depth.

$$p = \frac{\partial^2}{\partial z^2} M(z) \qquad [2]$$

The single pile was instrumented with five pairs of strain gauges along its length, as shown in Fig. 2. The discrete measurements of bending moments along the pile were interpolated using a cubic spline interpolation technique. A cubic spline is perhaps the simplest interpolation of discrete values that can be double differentiated (Wilson, 1998); however since the spline fits every point exactly, the interpolation is affected by high frequencies (dynamic component) upon differentiation. Therefore, the dynamic component of the bending moment records was filtered out before obtaining the bending moment distributions. The distribution of lateral resistance (p) was obtained by double differentiating the interpolated bending moment distribution with respect to depth.

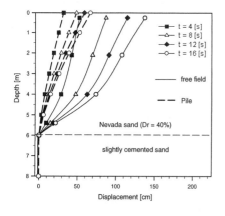

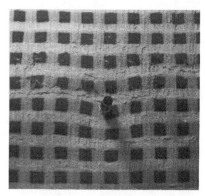

Figure 4: Profiles of free field and pile displacement

Figure 5: Pattern of soil displacement around the pile (z = 1 m)

Figure 7 shows profiles of lateral soil resistance (p) at different instants during the dynamic excitation. The lateral resistance varied between 5 and 20 kN/m below a depth of 2 m. However, near the surface, particularly at a depth of 1 m, the lateral resistance increased considerably during shaking, reaching values of about 50 kN/m.

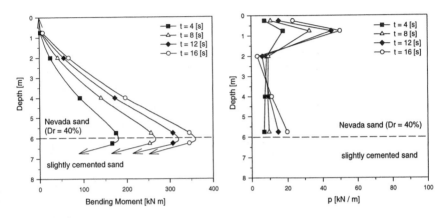

Figure 6: Profiles of bending moment   Figure 7: Profiles of lateral soil
                                                                        resistance

Most probably this negative excess pore pressure near the surface stiffened the soil close to the pile, enabling it to maintain a strong "grip" near the pile head which would explain the high pile displacement and bending moment, as well as the lack of pile rebound. This is consistent with the measured response of the single pile of model 1x1-w, saturated with water (González et al., 2005), where there were no negative excess pore pressures and the pile deformation and bending moments were much smaller. All this highlight a very important effect of the liquefied soil permeability on the pile response subjected to lateral spreading, depending if the dilative effect of the dynamic load on the soil, as well as the decrease in soil lateral pressure downstream of the pile, are able to be canceled out on time by the pore fluid coming from the free field to avoid the development of negative excess pore pressures.

CONCLUSIONS

Liquefaction-induced lateral spreading continues to be a major cause of damage to deep foundations. Currently there is a huge uncertainty associated with the maximum lateral pressures and forces applied by the liquefied soil, which translates into a similar huge uncertainty in the calculated maximum pile bending moments. This article presents results and analysis of one of the centrifuge tests that were conducted at the 150 g-ton RPI centrifuge to investigate the effect of soil permeability in the response of single piles and pile groups to lateral spreading. The most relevant conclusions are:

- In the model saturated with viscous fluid (model 1x1-v) the pile does not bounce back during base excitation, and bending moments and lateral displacement are much larger than the ones observed in the model saturated with water (model 1x1-w).

- The decrease in lateral stress on the downslope side of the pile, and large shear strains around the pile with an undrained response, seams to have been responsible for the development of negative excess pore pressure near the pile in the model saturated with viscous fluid, particularly near the ground surface.

- This decrease in pore pressure near the pile in model 1x1-v caused a large increase in the liquefied soil resistance around the pile, enabling it to maintain a strong "grip" near the pile head. This large lateral force sustained by the shallow liquefied soil over the pile head until the end of the excitation would explain why the bending moments and pile displacement were much larger than the ones in the model saturated with water.

- The use of colored sand was very useful to visualize the large area of influence around the pile in the model saturated with viscous fluid, which is consistent with the increase in force on the foundation in this case.

- In conclusion, the liquefied soil permeability is an extremely important and poorly understood factor over the deep foundation response subjected to lateral spreading. The phenomenon is complex, including the dilatancy tendency of soil to shear, the decrease in lateral pressure on the downslope side of the pile, and the time the fluid needs to flow from the free field in order to dissipate on time the negative excess pore pressure developed during the excitation near the pile. This study and discussion of model 1x1-v, as well as the comparisons with model 1x1-w and the other models discussed by González et al. (2005) suggest that the pile bending moments and lateral displacements in a silty sand in the field during an earthquake may also be much greater than in a clean sand.

REFERENCES

Abdoun, T., Dobry, R., O'Rourke, T. D., and Goh, S. H. (2003). "Pile response to lateral spreads: centrifuge modeling." *Geotechnical & Geoenvironmental Engineering*, Vol. 129, No. 10, pp. 869-878.

Berrill, J. B., Christensen, S. A., Keenan, R. J., Okada, W., and Pettinga, J. K. (1997). "Lateral-spreading loads on a piled bridge foundation." *Seismic Behavior of Ground and Geotechnical Structures*, (Seco, E. Pinto, ed.), Balkema, Rotterdam, pp. 173-183.

De Alba, P., and Ballestero, T. P. (2004). "Residual strength after liquefaction: a rheological approach." *Proc., 11th International Conference on Soil Dynamics and Earthquake Engineering*, Vol. 2, pp. 513-520.

Dobry, R., and Abdoun, T. (2001). "Recent studies on seismic centrifuge modeling of liquefaction and its effect on deep foundation." State-of-the-Art Paper, *Proc., 4th International Conf. on Recent Advances in Geotechnical Earthquake Engineering and Soil Dynamics* (Prakash, S., ed.), Paper SOAP 3, San Diego, CA, March 26-31, Vol. 2, 30 pages.

Dungca, J. R., Kuwano, J., Saruwatari, T., Izawa, J., Suzuki, H., and Tokimatsu, K. (2004). "Shaking table tests on the lateral response of a pile buried in liquefied sand." *Proc., 11th International Conference on Soil Dynamics and Earthquake Engineering*, Vol. 2, pp. 471-477.

González, L., Abdoun, T., and Dobry, R. (2005). "Effect of soil permeability on centrifuge modeling of pile response to lateral spreading." *Geotechnical & Geoenvironmental Engineering*, ASCE; (in preparation).

Hamada, M., Yasuda, S., Isoyama, R., and Emoto, K. (1986). "Study on liquefaction induced permanent ground displacements." *Research Rept., Assn. For Development of Earthquake Prediction*, Japan, November, 87 pages.

Hamada, M., and O'Rourke, T.D. (eds.) (1992). "Case studies of liquefaction and lifeline performance during past earthquakes." *Vol. 1: Japanese Case Studies*, National Center for Earthquake Engineering Research, SUNY-Buffalo, Buffalo, NY *(Tech. Rept. NCEER-92-0001*, February).

Hamada, M. (1998). "A study on ground displacement caused by soil liquefaction." *Proc., Japanese Society of Civil Engineers*, No. 596, Vol. III-43, pp. 189-208.

Higuchi, H., and Matsuda, T. (2002). "Effects of liquefaction-induced lateral flow of ground against a pile foundation." *Proc., International Conf. of Physical Modeling in Geotechnics*, St. John's Canada, Balkema, pp.465-470.

Hwang, J. I., Kim, C. Y., Chung, C. K., and Kim, M. M. (2004). "Behavior of a single pile subjected to flow of liquefied soil of an infinite slope." *Proc., 11th International Conference on Soil Dynamics and Earthquake Engineering*, Vol. 2, pp. 573-580.

Ishihara, K., Yasuda, S., and Nagase, H., (1996). "Soil characteristics and ground damage." *Soils and Foundations, 109-118,* January.

McCulloch, D.S., and Bonilla, M.G. (1970). "Effects of the earthquake of March 27, 1964 on the Alaska railroad." *Professional Paper* 545-D, U.S. Geological Survey.

Mizuno, H. (1987). "Pile damage during earthquakes in Japan (1923-1983)." *Proc., Session on Dynamic Response of Pile Foundations* (Nogami, T., ed.), ASCE, Atlantic City, April 27, pp. 53-77.

O'Rourke, T. D., and Hamada, M. (eds.) (1992). "Case studies of liquefaction and lifeline performance during past earthquakes, Vol. 2: United States Case Studies." *National Center for Earthquake Engineering Research*, SUNY-Buffalo, NY (Tech. Rept. NCEER-92-0002).

Swan, S.W., Flores, P.J., and Hooper, J.D. (1996). "The Manzanillo Mexico earthquake of October 9, 1995." *NCEES Bulletin, The Quarterly Publication of NCEER*, 10 (1), January.

Tokimatsu, K., Mizuno, H., and Kakurai, M. (1996). "Building damage associated with geotechnical problems." *Soils and Foundations,* 219-234, January.

Tokimatsu, K., (1999). "Performance of pile foundations in laterally spreading soils." *Proc., 2nd Intl. Conf. on Earthquake Geotechnical Engineering* (P. Seco e Pinto, ed.), Lisbon, Portugal, Vol. 3, pp. 957-964.

Wilson, D. W. (1998). "Soil-pile-superstructure interaction in liquefying sand and soft clay." *Ph.D. Thesis*, University of California at Davis, Davis, CA, USA.

Yokoyama, K., Tamura, K., and Matsuo, O. (1997). "Design methods of bridge foundations against soil liquefaction and liquefaction-induced ground flow." *Proc., 2nd Italy-Japan Workshop on Seismic Design and Retrofit of Bridges*, Rome, Italy, pp. 109-131.

Youd, T.L., (1993). "Liquefaction-induced damage to bridges." *Transportation Research Record,* published by the Transportation Research Board and the National Research Council, Washington, D.C., USA, 1411, 35-41.

COMPARISON OF CURRENT JAPANESE DESIGN SPECIFICATIONS FOR PILE FOUNDATIONS IN LIQUEFIABLE AND LATERALLY SPREADING GROUND

Akihiko Uchida[1], Kohji Tokimatsu, Member, ASCE[2]

ABSTRACT

Three design specifications (highway bridges, railway facilities, and building foundations) currently used for pile foundations in liquefiable and laterally spreading ground in Japan are described and compared in this paper. Although all the specifications take into account the effects of ground deformation due to liquefaction or lateral spreading on the performance of the pile foundation, the coefficient of subgrade reaction used in the three specifications varies considerably. The effect of the reduction of the coefficient of subgrade reaction due to liquefaction on the response of the pile is investigated for two types of piles. Finally, a design chart of the pile foundation in liquefiable and laterally spreading ground studied in the research committee of the Japanese Geotechnical Society is introduced.

INTRODUCTION

Ground deformation due to soil liquefaction or lateral spreading that occurred in the coastal area of Kobe city during the Great Hanshin earthquake of 1995 had significant influence on the damage to piles, indicating that the effects of ground deformation should be properly taken into account in design specification of piles. For this reason, many design specifications in Japan have been reformed and seismic deformation method has been adopted for the design works of piles. Because of lack of sufficient data for predicting the effects of ground deformation due to liquefaction or lateral spreading, the external force from ground displacement is different among

---

[1] Takenaka Corporation, Chiba, Japan.
[2] Tokyo Institute of Technology, Tokyo, Japan.

specifications.   In this paper, three typical design specifications (Highway Bridges, Railway Facilities and Building Foundations) in Japan are introduced to compare the view of a design of piles in liquefiable or laterally spreading ground, and their common features and differences are described.   Finally, a design chart of the pile foundation in liquefiable and laterally spreading ground summarized by the research committee of the Japanese Geotechnical Society is presented.

FUNDAMENTAL VIEW OF SPECIFICATIONS

The fundamental design procedure of the pile foundation in liquefiable and laterally spreading ground shown in Table 1 is summarized below for the three specifications:

(1) Specifications for Highway Bridges (JRA reformed in 2002)

In liquefied level ground, the design of the pile foundation against lateral load is performed by the seismic coefficient method with coefficient of subgrade reaction that reflects the effects of soil liquefaction, but the influence of the ground deformation due to liquefaction is not taken into consideration.   The reduction of coefficient of subgrade reaction due to liquefaction is defined as a function of the FL value (safety factor against liquefaction), a liquefaction strength ratio, the depth of ground, and the earthquake level, as shown in Table 2.

In laterally spreading ground, on the other hand, the earth pressure is applied on the pile foundation without inertial force.   The earth pressure used for design is about 0.3 times the total vertical stress in the liquefiable layer, and the passive earth pressure in a non-liquefied surface layer.

(2) Design Standard for Railway Facilities (RTRI reformed in 1999)

The design of the pile foundation for the lateral load is performed mainly with the inertia force of superstructure. Although the ground deformation due to liquefaction is not considered in the design, the additional inertia force associated with the grade of liquefaction is applied on the pile instead of the ground deformation. The subgrade reaction (the coefficient of subgrade reaction) is reduced as a function of both FL value and the depth for the liquefiable layer, as shown in Fig1.

The seismic deformation method is used for the design of piles in laterally spreading ground. The displacement due to lateral spreading is estimated with consideration of both the thickness of liquefiable layer and the movements of quay wall, and the ground displacement is applied on the pile foundation through the subgrade reaction. The subgrade reaction used for laterally spreading layer is 1/1000 of the initial value, which is different from that used for the liquefiable layer.

(3) Recommendations for Design of Building Foundations (AIJ reformed in 2001)

The design of the pile foundation is conducted by the seismic deformation method, which considers the inertia force of superstructure and the ground deformation. The ground deformation due to liquefaction manifested through the accumulated shear strain in the depth direction for liquefiable layer with the Na-value is shown in Fig2. The subgrade reaction is reduced with both the Na-value and the depth for the liquefiable layer, as shown in Fig3. The subgrade reaction used for the inertia force of the superstructure in liquefiable layer is the same value as that used for the ground deformation due to liquefaction.

The ground deformation due to lateral spreading is estimated with both the thickness of liquefiable layer and the movements of quay wall, and it acts on the pile foundation through the subgrade reaction. There is no description on the reduction of the subgrade reaction for the lateral spreading, and it is considered to be the same value as for liquefaction.

SUBGRADE REACTION USED IN SPECIFICATIONS

Three design specifications introduced in this study use subgrade reactions (coefficients of subgrade reaction) to resist the inertial force of the superstructure. Different methods of estimating the subgrade reaction are used in three specifications. The value of the coefficient of subgrade reaction calculated from N value for the sand layer was compared. Fig4 shows the coefficient of subgrade reaction, which is estimated by the following formula for three specifications. The estimated coefficients of subgrade reaction vary, with the largest kh in JRA specification, about 3 times that of AIJ.

<Common condition>
Type of soil: sandy ground
Young modulus, E_0: estimate from N value
Type of pile: cast in place concrete pile (single pile)

(1) Specifications for Highway Bridges (JRA reformed in 2002)
The coefficient of subgrade reaction, k_h, in JRA can be determined by the following formulas.

$$k_h = k_{h0}(B_H/0.3)^{-3/4} \quad (kN/m^3) \quad (1)$$
$$B_H = \sqrt{(D/\beta)} \quad (2)$$
$$\beta = (k_h D/4EI)^{1/4} \quad (3)$$
$$k_{h0} = \alpha E_0/0.3 \quad (kN/m^3) \quad (4)$$
$$E_0 = 2800N \quad (kN/m^2) \quad (5)$$

Where,
B_H : normalized width of pile (m)
D : pile diameter (m)
β : characteristic value
EI : bending rigidity of pile (kNm^2)

α : constant value (α =2 for Eo evaluating from N-value)

E_0 : Young's modulus (kN/m^2)

N : SPT N-value

(2) Design Standard for Railway Facilities (RTRI reformed in 1999)

The coefficient of subgrade reaction, k_h, in RTRI can be estimated by the following formulas.

$$k_h=f_{rk}(0.6 \alpha E_0 D^{-3/4}) \qquad (6)$$
$$E_0=2500N \qquad (kN/m^2) \qquad (7)$$

Where,

f_{rk} : resisting factor of soil (=1.0)

α : constant value (α =2 for Eo evaluating from N-value)

D : pile diameter (m)

E_0 : Young's modulus (kN/m^2)

(3) Recommendations for Design of Building Foundations (AIJ reformed in 2001)

The coefficient of subgrade reaction, k_h, for AIJ can be evaluated by the following formulas.

$$k_h=k_{h0}y^{-1/2} \qquad (kN/m^3) \qquad (8)$$
$$k_{h0}= \alpha \xi E_0 B^{-3/4} \qquad (kN/m^3) \qquad (9)$$
$$E_0=700N \qquad (kN/m^2) \qquad (10)$$

Where,

y : lateral displacement of pile (cm)

α : constant value (m^{-1}) (α =80 for Eo evaluating from N-value in sandy soil)

ξ : constant for group piles (1.0 for single pile)

E_0 : Young's modulus (kN/m^2)

B : width of pile (cm)

EFFECT OF SUBGRADE REACTION ON PILE RESPONSE

In order to evaluate the effect of the subgrade reaction (the coefficient of subgrade reaction) on the pile response in liquefiable layer, the displacement and moment distributions in pile against inertial force were computed with non-reduced and reduced subgrade reactions. One is PC pile, which has small rigidity against lateral load, the other is RC pile, which has large rigidity. The target piles were the ones that suffered damage due to liquefaction in the Great Hanshin Earthquake. The coefficient of subgrade reaction in this study was estimated by using the Recommendations for Design of Building Foundations (AIJ reformed in 2001).

1) Prestressed concrete pile (PC pile)

Fig5 shows the soil profile and the pile condition used in this calculation. The superstructure was a three-story building constructed in 1973.   Several large horizontal cracks occurred at a depth of about 9 meters. Since the reclaimed fill layer in surface had a high potential for liquefaction, the shear and moment distributions in pile were determined with and without considering the reduction of the subgrade reaction of the liquefiable layer. The axial load was 201kN, which was estimated for the extracting side of piles in the earthquake. The inertia force acting on the pile head was estimated to be 120kN, which was 0.13 at the axial load of steady state.

Fig6 shows the distribution of the bending moment and the shear stress of the pile. The solid lines in the figure show the allowable bending moment and the allowable shear stress of the pile, respectively. It was found that the response in liquefiable layer is larger in case for reduction of the subgrade reaction. The large bending moment calculated in this study is close to the allowable value of the pile. Because of the small rigidity of the PC pile, the subgrade reaction might affect the pile response.

2) Cast in place concrete pile (RC pile)

Fig7 shows the soil profile and the pile condition used in this calculation. The superstructure was the highway bridge constructed in 1991. The reclaimed fill layer in surface was a high potential against liquefaction, and slight horizontal cracks were detected by the survey from the pile head to a depth of about 5 meters in the earthquake. The moment and shear force distributions in pile were computed with and without reduction of subgrade reaction. The axial load was 908kN, which was estimated for the extracting side of piles in the earthquake. The inertia force acting on the pile head was estimated as 516kN, which was 20% of the axial load in steady state.

Fig8 shows the distribution of the bending moment and the shear stress of the pile. The moment and shear force distributions in the RC pile computed with the two different subgrade reaction are almost the same. Because of the large rigidity of the RC pile, it is thought that the subgrade reaction should not affect the pile response.

The results of the above two examples suggest that the evaluation of subgrade reaction might be important for small rigidity piles in design works.

DESIGN CHART FOR PILE FOUNDATION IN LIQUEFIABLE AND LATERALLY SPREADING GROUND

Although details of the pile foundation design are different among specifications currently used in Japan, they have some common views for the design process. The research committee (the committee of the pile foundation design in liquefiable ground) of the Japanese Geotechnical Society, developed the design chart shown in Figs9 and 10 for the pile foundation in liquefiable and laterally spreading ground, which was referred from typical specifications in Japan.

In the liquefied soil, after evaluating the liquefaction potential of the ground and judging whether the effect of liquefaction is taken into consideration in the design of piles, the pile stresses should be calculated against the inertia force of the superstructure and ground deformation due to liquefaction. The computed pile stresses are then compared with the acceptable values of the piles.

In laterally spreading ground, the response of the pile can be evaluated against the ground deformation or the earth pressure due to lateral spreading. The effect of the inertia force of superstructure may be neglected in laterally spreading ground.

It is believed that the design chart shown here will be effective for making a unification standard of Japan in liquefiable and laterally spreading ground.

DISCUSSIONS

Some of the Japanese specifications consider the ground deformation due to liquefaction and lateral spreading for the pile foundation design. Although the pile response is significantly affected by the ground deformation, the methods for estimating ground deformation including the coefficient of subgrade reaction are different among the specifications. The coefficient of subgrade reaction and its reduction due to liquefaction therefore should be the peculiar value corresponding to the soil property in geotechnical aspect.

Moreover, the ground deformation and the earth pressure due to liquefaction or lateral spreading should be estimated by considering the soil properties. Thus, it is suggested that the Japanese Geotechnical Society should choose the best way for geotechnical design values such as the coefficient of subgrade reaction and the ground deformation due to liquefaction.

ACKNOWLEDGEMENTS

This study is quoted from the result performed by the research committee (the committee of a pile foundation design in liquefiable ground) of the Japanese Geotechnical Society. The authors express their gratitude to the members of the research committee.

REFERENCES

The Research Committee of JGS (2004). Proceedings of the symposium for a pile foundation design in liquefiable ground. (in Japanese)

Japan Road Association (JRA) (2002). Specifications for highway bridges, part V: seismic design. (in Japanese)

Railway Technical Research Institute (RTRI) (1999). Design standard for railway facilities– seismic design. (in Japanese)

Architectural Institute of Japan (AIJ) (2001). Recommendations for design of building foundations. (in Japanese)

Table 1: Fundamental view of pile foundation design

		Highway Bridges (JRA)	Railway Facilities (RTRI)	Building Foundations (AIJ)
Liquefaction	Inertia force	← P	← P	← P
	Ground movement force	Not consider	← P+ α	deformation y_G
	Subgrade reaction	Reduce with · FL-value · Depth · Liquefaction strength · Earthquake Level	Reduce with · FL-value · Depth	Reduce with · Na-value · Depth
Lateral spreading	Inertia force	Not consider	Not consider	← P(liq)
	Ground movement force	Non-liq q_{NL} / Liq q_L	deformation y_G	deformation y_G
	Subgrade reaction	×	Liq.layer:1/1000 Non-liq.layer:No reduction	Same as liquefaction

Table 2: Reduction factor for Highway Bridges

F_L	Depth	Liquefaction strength			
		$R \leqq 0.3$		$0.3 < R$	
		Level 1	Level 2	Level 1	Level 2
$F_L \leqq 1/3$	$0 \leqq x \leqq 10$	1/6	0	1/3	1/6
	$10 < x \leqq 20$	2/3	1/3	2/3	1/3
$1/3 < F_L \leqq 2/3$	$0 \leqq x \leqq 10$	2/3	1/3	1	2/3
	$10 < x \leqq 20$	1	2/3	1	2/3
$2/3 < F_L \leqq 1$	$0 \leqq x \leqq 10$	1	2/3	1	1
	$10 < x \leqq 20$	1	1	1	1

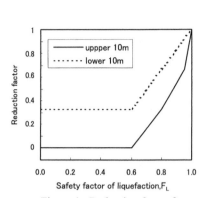

Figure 1: Reduction factor for
Railway Facilities

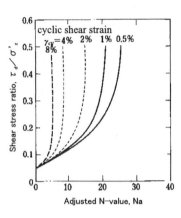

Figure 2: Estimating shear strain in
liquefiable layer

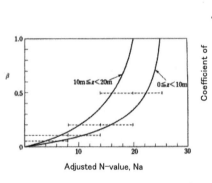

Figure 3: Reduction factor for Building
Foundations

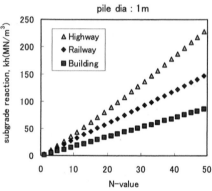

Figure 4: Comparison of coefficient of
subgrade reaction

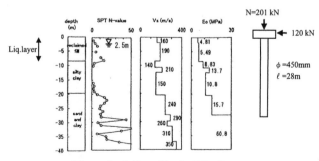

Figure 5: Soil profile for PC pile

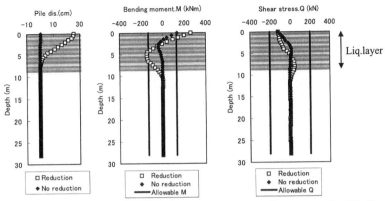

Figure 6: Effect of reduction of subgrade reaction on pile response for PC pile

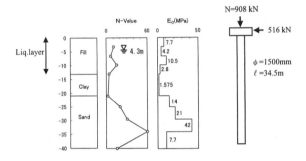

Figure 7: Soil profile for RC pile

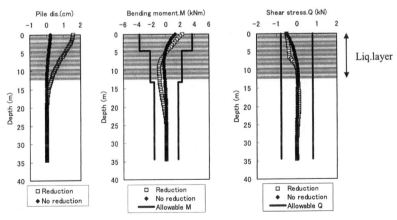

Figure 8: Effect of reduction of subgrade reaction on pile response for RC pile

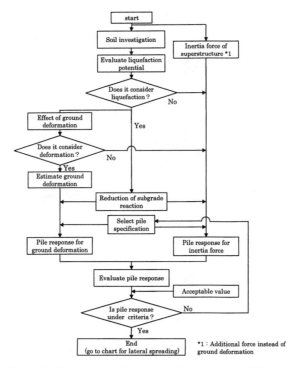

Figure 9: Design chart for pile foundation in liquefiable ground

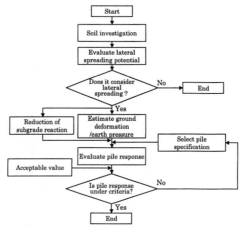

Figure 10: Design chart for pile foundation in laterally spreading ground

LATERAL SPREADING FORCES ON BRIDGE PILES

Jin-xing Zha, Ph.D., P.E.[1]

ABSTRACT

This paper presents a simplified procedure for computing liquefaction–induced lateral spreading forces on piles or abutment walls supported on piles. This approach is in essence to develop two curves of load versus displacement behavior to consider the soil-pile-structure interaction in a "decoupling" way. A curve of lateral spreading force versus displacement is first established by slope stability analysis for the soil mass behind the wall based on the Newmark chart that relates the standardized maximum displacement of the soil mass to the limiting acceleration coefficient. A second curve of lateral load versus displacement for the abutment wall supported on piles is then obtained by push analysis of the wall structure under the lateral spreading force. The intersection between the two load-displacement curves is considered as the lateral spreading force on the abutment wall. Results of analyses performed for one bridge are presented in this paper to illustrate the proposed method. It appears that the simplified method of accounting for the pile-wall-soil interaction can satisfactorily predict the lateral spreading forces on abutment walls supported by piles without the uncertainty in determining the seismic coefficient related to the use of the Mononobe-Okabe method.

INTRODUCTION

There are two typical cases in which lateral spreading forces on piles need to be provided to bridge engineers for the purpose of structural design. The first case is for piles at abutments where the embankment slope is in unstable condition due to liquefaction in underlying soil layers during earthquake shaking and tends to slide

---

[1] Geotechnical Services, California Department of Transportation, 5900 Folsom Boulevard, Sacramento, CA 95819, U.S.A.

downward. Consequently, the soil mass behind the wall imposes lateral spreading forces on the abutment walls/piles as shown in Figure 1. In the second case, piles at bents are subject to lateral spreading forces from the non-liquefiable top crust overlying liquefiable layers. If the thickness of the top crust is limited, say smaller than 3 meters (10 ft), the Rankine formula of passive soil pressure is utilized to estimate the lateral spreading forces on piles at bents as shown in Figure 2. When the thickness is large, engineering judgment is required to assess whether the passive soil pressure is too high to be applied since it increases dramatically with the thickness. In this aspect, investigation is to be carried out to see whether a displacement-based method like the approach that is to be presented for abutment piles can be used. Thus, in the following sections only discussed is the method for calculating lateral spreading forces on piles at abutments as in the first case.

For a proposed bridge project, site liquefaction evaluation is performed using the simplified Seed method (Seed and Idriss 1971 and Youd and Idriss 1997), based on the information of site seismicity from the Caltrans 1996 California Seismic Hazard Map – a GIS database of seismicity for California State and Counties bridges. After liquefiable soil layers at the site are identified, the potential for lateral spreading at abutments due to slope instability during earthquake shaking is examined by performing pseudostatic analysis of slope stability using residual undrained shear strengths for liquefied layers. If the abutment slope is susceptible to lateral spreading, the lateral spreading force on piles is to be estimated with a simplified procedure.

For seismic design of piles at bridge abutments or retaining walls, the Mononobe-Okabe method may be used to determine the soil pressures including dynamic inertia effect by applying assumed pseudo-static horizontal and/or vertical accelerations. To account for the effect of lateral displacement of a retaining structure on the seismic soil pressure, different intensities of earthquake shaking in terms of a seismic coefficient are assigned to apply to retaining structures with different restraints that limit lateral displacement. According to the Standard Specifications for Highway Bridges (AASHTO 1996), a seismic coefficient of fifty percent of the designed peak horizontal ground acceleration is recommended for bridge abutments or retaining walls that may undergo significant horizontal displacement and a seismic coefficient equal to three times the value for walls without restraints is assumed for abutments or retaining walls that are restrained from horizontal displacement by anchors or piles. One shortcoming with the use of the Mononobe-Okabe method is the uncertainty in determining the seismic coefficient according to the restraints that prevent the wall from freely moving.

Alternatively, a "trial and error" procedure was used years ago by designers (Perez-Cobo and Abghari 1996) with Caltrans for estimating liquefaction-induced lateral spreading force on piles for the Salinas River Bridge on State Highway 101 in Monterey County, California. For that project, the Newmark chart was utilized to determine the maximum displacement of a sliding soil mass subject to seismic shaking specified by an assumed yield acceleration and the lateral spreading force on piles by the sliding soil was calculated with p-y soil spring models based on the maximum displacement. Stability analysis of the sliding soil slope was performed to find the minimum pile resistance required to maintain the slope stability. Iterative calculation was conducted for different assumed yield accelerations until the minimum pile resistance matches the lateral spreading force on piles from p-y modeling. This concept with variations in details has since been used for other Caltrans projects, including the Oakland Mole SFOBB East Span Seismic Safety project (Law 2000). However, the p-y type procedure for estimating lateral spreading force on piles (not retaining walls or bridge abutments) overestimates the force since the maximum displacement of the sliding soil mass is larger than the relative displacement between the pile and the soil. In addition, the displacement compatibility between the pile deflection and the movement of the sliding soil is not maintained in the analysis by Law (2000), as it was stated in the report that the pile deflection would be different from the prescribed displacement profile for the sliding soil mass by a relative displacement between two ends of the p-y springs that impose load on the pile.

Martin et al. (2002) recommended a design approach for calculating liquefaction induced lateral spreading force on piles for bridges subject to the Maximum Considered Earthquake (MCE) with a recommended design return period of 2475 years (probability of exceedance of 3% in 75 years), which is a significant increase from the current AASHTO (1996) recommended return period of 475 years that corresponds approximately to a probability of exceedance of 15% in 75 years. This method is conceptually similar to the above-discussed "trial and error" procedure and is presented in detail with a design flow chart and a 13-step procedure followed by two case studies. This procedure involves much integration of estimating lateral spreading force on piles into structure design process such as sizing the pile prior to determination of the lateral pile load demand and assessing formation of a plastic hinge in a pile. Consequently a great effort of cooperation between geotechnical and bridge engineers is required to accomplish the calculation of lateral spreading force. Hence, for engineers without specific training, this 13-step procedure is rather hard to be followed for estimating the lateral spreading force on piles.

Zha (2004) presented a simplified two-curve procedure for computing lateral spreading force on bridge abutment walls supported on piles due to slope instability during earthquake shaking. It utilizes the Newmark chart to correlates the maximum displacement of the sliding block/soil mass with the limiting acceleration coefficient. This approach is conceptually an extension of the design method suggested by Richards and Elms (1979) for analyzing seismic behavior of gravity retaining walls with a specified limiting wall displacement to which the limiting acceleration coefficient is correlated (Franklin and Chang 1977). In the analysis, the pile-wall-soil interaction is considered in a "decoupling" manner in which two separate load versus displacement curves are developed for the sliding block/soil mass from slope stability analysis and for the retaining wall/pile system from push analysis, respectively. The solution for the interaction between the sliding soil mass and the retaining pile/wall system is then obtained by combining the two individual load versus displacement curves and finding the intersection of them.

In this paper, the simplified two-curve procedure is first briefly outlined and a case study is then presented to illustrate the procedure. More detail of the simplified approach for estimating lateral spreading force on piles at abutments due to slope instability during earthquake shaking can be found in the paper by Zha (2004).

SIMPLIFIED PROCEDURE

The simplified procedure that is utilized for computing lateral spreading forces on bridge abutment walls supported on piles due to slope instability during earthquake shaking is in essence to develop two curves of load versus displacement behavior to consider the soil-pile-structure interaction in a "decoupling" way. A curve of lateral spreading force versus displacement is first established by slope stability analysis for the soil mass behind the wall based on the Newmark chart that relates the standardized maximum displacement of the soil mass to the limiting acceleration coefficient. A second curve of lateral load versus displacement for the abutment wall supported on piles is then obtained by push analysis of the wall structure under the lateral spreading force. The intersection between the two load-displacement curves is considered as the lateral spreading force on the abutment wall in the sense that equal loads are applied to the sliding soil mass and the pile/wall retaining system and the displacement compatibility is maintained between the pile and the soil.

Pseudostatic Slope Stability Analysis

Much work with generation of the first load – displacement curve is the pseudostatic slope stability analysis based on the simplified Janbu method. By this method, the factor of safety (FOS) against shear failure is defined as:

$$FOS = F_R / F_D \qquad (1)$$

in which, F_R and F_D are respectively the horizontal component of the soil resistance along the failure surface and the horizontal driving force of the sliding block mass. The factor of safety (FOS_{Reinf}) against shear failure for the slope reinforced with piles may be expressed as:

$$FOS_{Reinf} = (F_{Pile} + F_R)/ F_D \qquad (2)$$

in which, F_{Pile} is the pile lateral resistance required to maintain the slope stability with piles for a designated factor of safety, FOS_{Reinf}. By combining Equations (1) and (2), one can get:

$$F_{Pile} = F_R (FOS_{Reinf}/FOS - 1) \qquad (3)$$

For a given seismic acceleration coefficient (in unit of g, gravity acceleration), a pseudostatic analysis of slope stability based on the simplified Janbu method generates the safety factor (FOS) and the horizontal component (F_R) of the soil resistance along the failure surface. Consequently, the required pile lateral resistance can be obtained from Equation (3) to maintain the slope stability with a designated factor of safety, FOS_{Reinf}.

Two Load – Displacement Curves

The first curve for the sliding block/soil mass can be constructed by performing the above stated slope stability analysis with input as given in Table 1. For each of the limiting accelerations listed in Table 1 (Franklin and Chang, 1977), depending on the magnitude of earthquake, a required pile lateral resistance calculated from Equation (3) and the corresponding permanent displacement in Column 2 make one point on the curve of lateral spreading force vs. displacement. For more accuracy in analysis or for "rare" earthquake events, such as those with a recommended design return period of 2475 years, the curves of sliding displacement versus limiting acceleration coefficient as given in Table 1 should be replaced with results from site-specific time history analysis based on the Newmark sliding block integration (Newmark 1965).

Table 1. **Mean Permanent Displacement of Slide Block as a Function of k/A* for Different Magnitudes of Earthquakes (Soil Sites)**

No.	Permanent Displacement mm (inches)	Magnitude of Earthquake		
		5.2-6.0	6.5	7.7
1	1016 (40)	0.14	0.17	0.2
2	178 (7)	0.34	0.37	0.41
3	76 (3)	0.44	0.47	0.51
4	25.4 (1)	0.54	0.6	0.62
5	7.6 (0.3)	0.64	0.7	0.74
6	2.5 (0.1)	0.74	0.79	0.83
7	0.25 (0.01)	1.0	1.0	1.0

* k/A is the ratio of limiting acceleration coefficient to peak ground acceleration coefficient.

Development of lateral load – deflection curve – Curve 2 as shown in Figure 5 for the bridge abutment wall supported on piles may be completed by push analysis of the global structure model to take advantage of the passive soil pressure behind the wall at the other abutment. In many cases, this can be simplified to a laterally loaded pile analysis if pile head connection is approximated using adequate spring restraints. The pile analysis is performed by incrementally applying the required pile lateral resistance or lateral spreading force obtained from Equation (3). The point of application for the lateral spreading force may be approximately at the mid-height of the sliding block/soil mass with an assumption that the spreading force is uniformly distributed between the pile and the sliding soil (Seed and Whitman 1970; Wood 1973; and Richards and Elms 1979). The point of load application at the mid-height of the wall appears to generate higher moment and shear on the pile cross-section than that at the bottom one third of the wall.

With the two load-displacement curves developed, the lateral spreading force on piles/walls is determined as the intersection of the two curves – Curves 1 and 2 as shown in Figure 5. The lateral spreading force – displacement curve for the sliding soil mass (Curve 1) typically (at least at four bridge sites for which this simplified method was applied to estimate lateral spreading forces) approaches an asymptotic value within a displacement of 380 mm (15 inches). An upper bound load value on the curve at a displacement ranging from 25 to 130 mm (1 to 5 inches), depending upon the ductility of the pile, may be defined as the lateral spreading force without the second curve. In the case in which Curve 2 as shown in Figure 5 is ignored, this procedure can be truly considered a decoupling method that can be used during initial design phase.

LAKE HODGES BRIDGE PROJECT

The proposed project located on California State Route 15 in San Diego County includes replacing the existing left and right bridges (existing Bridge No. 57-0040R/L). The new structure including the left, middle, and right bridges is a 215m (705 ft) long and 73m (240 ft) wide, 5-span PC/PS concrete box girder superstructure supported on the abutments and piers founded on Cast-In-place-Drilled-Hole (CIDH) piles.

Site Seismicity

According to the Caltrans California Seismic Hazard Map-1996, there are two active faults within 32 kilometers of the bridge site: the Newport-Inglewood-Rose Canyon/E (NIE, $M_w = 7$, style Strike-slip) fault on the southwest side and the Whittier-Elsinore (WEE, $M_w = 7.5$, style Strike-slip) fault on the northeast side. The distances between the bridge site and the two faults (NIE & WEE) are respectively 23.5 and 31.3 kilometers. The controlling fault for the bridge structure design is the NIE fault which is capable of generating a maximum credible earthquake of moment magnitude $M_w = 7$. The horizontal Peak Bedrock Acceleration at this site is estimated to be 0.2g (gravity acceleration) based on the 1996 Geomatrix attenuation relationship and to be 0.3g according to the Caltrans Seismic Hazard Map. For liquefaction evaluation and slope stability analysis, PBA = 0.2g and PGA = 0.27g are used.

Potential for Liquefaction and Lateral Spreading

From the Log of Test Borings completed in 1974 and 2002 for this bridge, there are at the abutments approximately 10 meters (33 ft) thick embankment fill materials from elevation +92.5 to 102.5 meters (+303 to +336 ft), consisting of dense to very dense sand with scattered gravel overlying loose to dense sandy silt. Subsurface materials at the middle of the channel mainly consist of top layers of very loose to compact sand with silt underlain by moderately decomposed granitic rock at elevation 82 meter (+268 ft). Groundwater was encountered at elevation +89.6 m (+294 ft) on December 19, 1974 during the drilling operation, which was at a depth of 1.8 meters (6 ft) below the original ground. Surface water can be observed during the wet season. Due to its fluctuation, a groundwater table at elevation +92.5 meters (+303 ft) is assumed for soil liquefaction analysis.

Based on the logs, the top 9.1 m (30 ft) thick layers of loose sand within the channel are susceptible to soil liquefaction during earthquake shaking. The vertical soil residual resistance of the liquefiable soil on piles can be considered negligible. The ground settlement due to soil liquefaction is estimated to be 50 mm (2 inches).

A study of the lateral spreading was conducted by performing slope stability analysis under seismic shaking using the program XSTABL. The potential for lateral spreading at the bridge site is high due to soil liquefaction at the toe and underneath the embankment fill at abutments. The embankment with a slope of approximately 2:1 (H:V) is in unstable condition and tends to slide downward the channel during earthquake shaking since there is not sufficient shear strength in the loose sand and liquefied soil at the toe of the slope. Tendency for the soil behind the abutment wall to move toward the lake channel will impose soil pressure on the abutment walls and the piles at Abutments 1 and 6. However, the sliding soil mass may not impose significant lateral spreading force on CIDH piles at Piers 2 through 5 as the top 9.1 m (30 ft) thick layers of loose sand surrounding the pier piles are liquefiable.

Lateral Spreading Force

The lateral spreading force on piles at Abutments 1 and 6 is calculated based on the progressive slope stability analysis in which a residual undrained shear strength of c_e = 7.2 kPa (150 psf) is assumed for the liquefiable layer per its SPT blow count. It is found from the XSTABL results (the potential sliding surfaces shown in Figure 3) that the soil slope in front of the abutment wall/piles fails during the designated earthquake shaking. In light of the possibility that the soil mass in front of the abutment wall slips down or slides away from the wall due to tension failure induced by liquefaction underneath, a deep slope cut in front of the wall is presumed for the analysis. As shown in the schematic of a slope cut for the progressive slope stability analysis in Figure 4, a slope cut of 10 meters (33 ft) height in front of the abutment wall may be assumed to analyze the slope stability behind the wall. A safety factor of 1.1 is assumed to obtain the force demand on the abutment wall/piles in order to maintain the slope stability under seismic shaking with an intensity of k = PGA = 0.27g. A safety factor greater than 1.0 should be designated since the lateral spreading force is not the only load acting on the piles. The difference between the two curves after the intersection as shown in Figure 5 may be viewed as the room for any uncertainty in determination of the total lateral load demand on a pile.

Following the steps discussed above, the first curve of lateral spreading force versus displacement for the soil slope with a cut of 10 meters (33 ft) height in front of the abutment wall as shown in Figure 3 is established as plotted in Figure 5. Calculation of lateral spreading forces on piles is given in detail in Table 2. The lateral spreading force on piles should be incorporated into the analysis of pile lateral load – deflection behavior by applying an approximately uniformly distributed pressure on the upper portion of the pile. However, push analysis for the second curve of pile lateral load – deflection behavior is not performed as the first curve approaches an asymptotic value within a displacement of 254mm (10 inches). Nevertheless, a second sample curve is plotted against the first one in Figure 5 for illustration only. From Figure 5, the lateral spreading force can be 1068kN (240 kips) per pile at Abutments 1 and 6 at a displacement of 55mm (2.2 inches) if an effective retaining width of 3.65m (12 ft) per pile is configured. The lateral spreading force is recommended for design of substructure and superstructure.

SUMMARY AND DISCUSSION

For piles at abutments where there is slope instability due to liquefaction in underlying soils, a displacement-based method for computing lateral spreading forces is proposed as a two-curve procedure based on the Newmark sliding block model. The procedure is simple to use since the pile-wall-soil interaction is considered in a "decoupling" manner. One case study of a bridge site is presented to illustrate the proposed method. It appears that the proposed method can satisfactorily predict the lateral spreading forces on piles or abutment walls supported by piles without the uncertainty in determining the seismic coefficient with the use of the conventional Mononobe-Okabe method. The proposed method may be applied to piles at bents when the thickness of the top crust undergoing lateral spreading is large.

During earthquake shaking, a structure initially responds elastically and reaches the maximum forces on its member cross-sections before any member and supporting soil foundation materials yield. Due to yielding, the stiffness of the soil-structure system softens and its structural period increases, which may lead to reduced earthquake loads in its members. At a liquefiable site, the reduced earthquake load may be accompanied by the lateral spreading force. In other words, since it is unlikely that the maximum inertial load and the lateral spreading force occur at the same time, the lateral spreading force (LSF) on piles may be used for structure design in two load combinations if earthquake loads control. One combination is the full lateral spreading force plus factored earthquake loads (EQ) transferred through abutment wall or bent columns (LSF + βEQ), and the other is the earthquake load combined with a factored lateral spreading force (EQ + βLSF), in which $\beta = 0 - 1$, determined from tests or by engineering judgment. In the case where the column plastic moment and shear govern the design, the full lateral spreading force should be combined with the column plastic moment and shear.

Although the Newmark sliding block model is widely acceptable for seismic slope stability analysis, verification of slope permanent deformations from test data and other numerical methods such as finite element model is desirable. It is a rational assumption that the sliding soil mass/block for embankments or dams undergoes only unrecoverable downward displacement under low frequency seismic excitations. However, it may overestimate the permanent displacement for slopes reinforced with piles that tend to elastically or elasto-plastically resist the downward movement of the soil mass/block.

ACKNOWLEDGMENTS

The author would like to thank Abbas Abghari, Angel Perez-Cobo, and Mahmoud Khojasteh, who all are with Caltrans, for their comments and suggestions throughout the preparation of this paper.

REFERENCES

AASHTO (American Association of the State Highway and Transportation Officials) (1996). *Standard Specifications for Highway Bridges*, 16th Edition

Franklin, A. G. and Chang, F. K. (1977). "Earthquake Resistance of Earth and Rock-Fill Dams." *Report 5: Permanent Displacements of Earth Embankments by Newmark Sliding Block Analysis, Miscellaneous Paper S-71-17*, Soils and Pavements Laboratory, U.S. Army Engineer Waterways Experiment Station, Vicksburg, Mississippi

Law, H. (2000). "Lateral Spreading on Pile at the Oakland Mole SFOBB East Span Seismic Safety Project." *Memorandum: 09-12-2000*, Earth Mechanics, Inc., Fountain Valley, California

Martin, G. R., March, M. L., Anderson, D. G., Mayes, R. L., and Power, M. S. (2002). "Recommended design approach for liquefaction induced lateral spreads." *Proceedings of the 3rd National Seismic Conference and Workshop on Bridges and Highways*, MCEER-02-SP04, Buffalo, New York

Newmark, N. M. (1965). "Effects of Earthquakes on Dams and Embankments." *Geotechnique*, London, England, Vol. 15, No. 2, pp.139-160

Perez-Cobo, A. and Abghari, A. (1996). "Lateral Spreading and Settlement Potential for Salinas River Bridge (Bridge No. 44-0002)." *Memorandum: 02-22-1996*, Caltrans-Office of Structural Foundations, Sacramento, California

Richards, R. and Elms, D. (1979). "Seismic Behavior of Gravity Retaining Walls." *Journal of Geotechnical Engineering*, ASCE, pp. 449-464

Seed, H. B. and Idriss, I. M. (1971). "Simplified Procedure for Evaluating Soil Liquefaction Potential." *Journal of Geotechnical Engineering*, ASCE, Vol. 97, pp.1249-1273

Seed, H. B. and Whitman, R. V. (1970). "Design of Earth Retaining Structures for Dynamic Loads." *Lateral Stresses in the Ground and Design of Earth Retaining Structures*, pp. 103-147

Wood, J. H. (1973). "Earthquake-Induced Soil Pressures on Structures." *Report No. EER: 73-05*, Earthquake Engineering Research Laboratory, California Institute of Technology, Pasadena, California

Youd, T. L. and Idriss, I. M. (1997). *Proceedings of the NCEER workshop on Evaluation of Liquefaction Resistance of Soils*, National Center for Earthquake Engineering Research, Technical Report NCEER-97-0022, 276p.

Zha, J. (2004). "Lateral Spreading Forces on Bridge Abutment Walls/Piles." *Geotechnical Special Publication No. 126*, ASCE, Vol. 2, pp.1711-1720

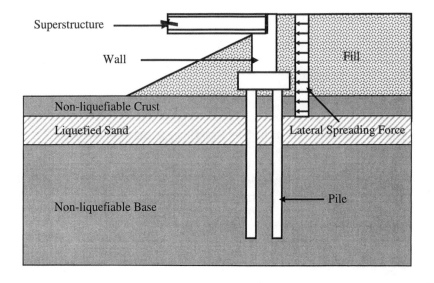

Figure 1. Lateral Spreading Force on Piles at Abutment

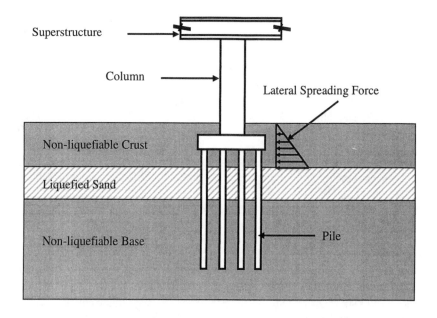

Figure 2. Lateral Spreading Force on Piles at Bent

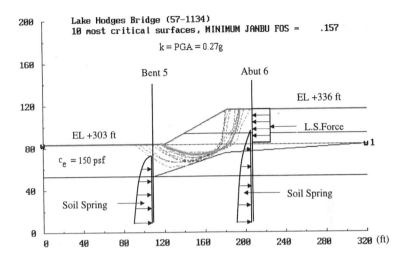

Figure 3. Lateral Spreading Forces on Piles and Abutment Wall

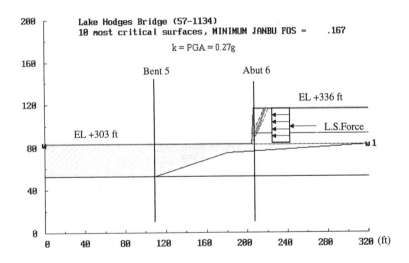

Figure 4. A Schematic of Slope Cut for Progressive Slope Stability Analysis

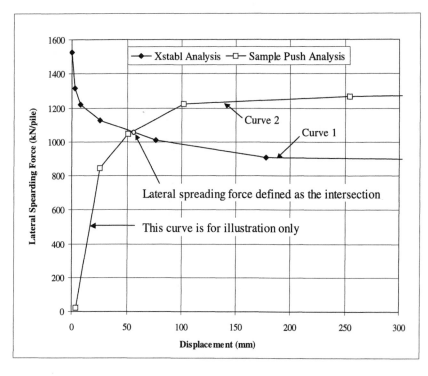

Figure 5. Lateral Spreading Force vs. Displacement of a Single Pile

Table 2. Calculation of Lateral Spreading Forces on a Single Pile

	Acceleration		Displacement	Factor of Safety	Resistance	Driving Force	Force Demand
No	a (g)	a/PGA	Δ (mm)	FOS	F_R (kN/m)	F_D (kN/m)	F_p (kN/pile)
1	0.05	18%	1016	0.235	62.9	268	847
2	0.11	40%	178	0.187	50.9	272	908
3	0.14	50%	76	0.174	51.9	298	1011
4	0.17	62%	25	0.164	54.1	330	1129
5	0.20	74%	7.6	0.176	63.6	361	1221
6	0.22	83%	2.5	0.165	63.6	385	1317
7	0.27	100%	0.25	0.176	79.4	451	1525
			PGA (g) =	0.27			
		Effective Pile Diameter (mm) =		1168			
	Effective Retaining Width per Pile (m) =			3.66			
	Factor of Safety for Slope Stability =			1.1			

SEISMIC EARTH PRESSURE ACTING ON EMBEDDED FOOTING BASED ON LARGE-SCALE SHAKING TABLE TESTS

Shuji Tamura[1] and Kohji Tokimatsu[2]

ABSTRACT

This paper examines earth pressure acting on an embedded footing and its effects on pile forces, based on both liquefaction and non-liquefaction tests using a large-scale laminar shear box. The following conclusions are drawn: (1) The total earth pressure defined by the difference in earth pressure between passive and active sides in the non-liquefaction tests varies significantly depending on its phase relative to the soil inertia around the embedded footing as well as on the relative displacement between soil and footing; (2) The total earth pressure in the liquefaction test, by contrast, depends mainly on the relative displacement because the soil inertia gets small in liquefied soil; (3) The total earth pressure in the non-liquefaction tests tends to be out of phase by 180 degrees with the superstructure inertia, reducing the shear force and bending moment at the pile head; and (4) The total earth pressure in the liquefaction tests tends to be in phase with the superstructure inertia, making the bending moment at the pile head large. A method for estimating the total earth pressure considering its phase relative to the superstructure inertia as well as the effects of soil inertia has been proposed. The proposed method gives a reasonable explanation of the difference in earth pressure between different tests.

INTRODUCTION

Extensive soil liquefaction that occurred on the reclaimed land areas of Kobe during the 1995 Hyogoken-Nambu earthquake caused vital damage to pile foundations. Over the past decade, a considerable number of studies have been made on the failure

---

[1] Disaster Prevention Research Institute, Kyoto University, Japan
[2] Tokyo Institute of Technology, Japan

mechanism of piles, based on the field investigation (Oh-Oka et al., 1998, Tokimatsu et al., 1998), centrifuge tests (Sato et al., 1995, Horikoshi et al., 1998) and numerical analyses (Miyamoto et al., 1997, Fujii et al., 1998). Many efforts have also been made on the evaluation of the horizontal subgrade reaction of piles (Wilson et al., 2000, Tokimatsu et al., 2002) as well as of the inertial and kinematic interaction of soil-pile-structure systems (Murono et al., 2000). In contrast, little attention has been given to dynamic earth pressure acting on an embedded footing and its effects on the failure mechanism of piles.

To investigate qualitatively the effects of inertial and kinematic forces, several series of large-scale shaking table tests were conducted on soil-pile-structure systems with dry sand (Tokimatsu et al. 2004) and with saturated sand (Tamura et al. 2000). Total earth pressure acting on the embedded footing was evaluated in the tests. The object of this paper is to examine influential factors affecting earth pressure acting on an embedded footing and then to propose a method for estimating the effects of the total earth pressure on the piles forces in both non-liquefaction and liquefaction tests.

SHAKING TABLE TESTS

This paper presents in details the earth pressure acting on an embedded footing based on the results of four shaking table tests, two with dry sand (cases DBL and DBS after Tokimatsu et al. 2004) and two with saturated sand (cases SBL and SBS after Tamura et al. 2000), as shown in Fig. 1. The tests were performed at NIED (National Research Institute for Earth Science and Disaster Prevention) in Tsukuba, Japan, using a large-scale laminar shear box 4.6 or 6.1 m high, 3.5 m wide and 12.0 m long (shaking direction) mounted on the shaking table. The soil used for cases DBL and DBS was Nikko Sand (e_{max}=0.96, e_{min}=0.57, D_{50}=0.31mm, F_c=5.4%). The relative density of the dry sand deposit was about 80%. The soil profile in cases SBL and SBS consists of two layers including a 4.5 m layer of Kasumigaura sand (e_{max}=0.98, e_{min}=0.65, D_{50}=0.42 mm) with a shear wave velocity of 90 m/s, which is underlain by a 1.5 m layer of gravel with V_S = 230 m/s. The water level was located at about GL-0.5 m. The relative density of the saturated sand deposit was about 35-50%.

A 2x2 steel pile was used for all the tests. The piles had a diameter of 16.5 cm, a thickness of 0.37 cm and a flexural rigidity EI of 1259 kNm^2. The pile heads were rigidly linked to the footing, while their tips were connected to the laminar shear box by hinges. The footing was modeled with a rigid steel box of 2.5 m (length) x 1.8 m (width) x 0.6 m (height) and embedded 0.5 m in the dry sand. The mass of the superstructure was 14200 kg and that of the footing was 2100 kg. The natural period of

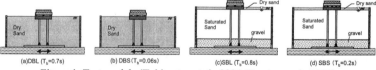

(a)DBL (T_b=0.7s)   (b) DBS (T_b=0.06s)   (c)SBL (T_b=0.8s)   (d) SBS (T_b=0.2s)

Figure 1: Test models (Tokimatsu et al., 2004, Tamura et al., 2000)

the superstructure under fixed footing conditions was about 0.7 s for case DBL and 0.06 s for case DBS. The former was longer than the natural period of the ground but the latter was shorter than that of the ground. The natural period of the superstructure in case SBL was about 0.8 s, which is longer than that of the ground before lique-faction but shorter than that after liquefaction. The natural period of the superstruc-ture in case SBS was about 0.2 s, which is shorter than that of the ground before and after liquefaction.

All the tests were excited by RINKAI92, which is a synthesized ground motion for the Tokyo Bay area. The amplitude of the motion was scaled to 240 cm/s^2. Accel-eration, displacement, excess pore water pressure and strain of the piles were re-corded during the tests.

NON-LIQUEFACTION SHAKING TABLE TESTS

Dynamic Response of Soil-Pile-Superstructure System

Figure 2 shows acceleration and displacement time histories of superstructure and ground surface, as well as those of input accelerations, relative displacement between footing and ground surface, total earth pressure and bending moment at the pile head in cases DBL and DBS (Tokimatsu et al. 2004). The displacements were calculated by the double integration of the accelerometer recordings. The total earth pressure, P can be evaluated by

$$P = Q - F \qquad (1)$$

in which Q = the sum of shear forces at the pile heads and F = the sum of the inertial forces of the superstructure and footing (Tamura et al. 2002a). The ground surface acceleration, footing and ground surface displacement time histories in case DBL are similar to those in case DBS, while both the magnitude and the predominant fre-quency of the superstructure acceleration, relative displacement, total earth pressure and bending moment in case DBL are different from those in case DBS.

The maximum bending moment in case DBS is about twice as large as that in case DBL. Although the maximum acceleration of the superstructure in case DBS is similar to that in case DBL, the maximum total earth pressure in case DBS is appar-ently smaller than that in case DBL. This suggests that the total earth pressure is the key controlling the difference in the bending moments between cases DBL and DBS.

To investigate the mechanism of the total earth pressure, the relations between the relative displacement and the total earth pressure during the tests are shown in Fig. 3. The relations between the two are elliptical in shape for both cases until 17 seconds. It is interesting to note that the total earth pressure in case DBS is smaller than that in case DBL, in spite of a larger relative displacement in case DBS. This indicates that the relative displacement may not be a primary factor causing the difference in total earth pressure between the two tests. The total earth pressure in case DBS tends to be stronger nonlinearity against relative displacement after 17 seconds, being smaller

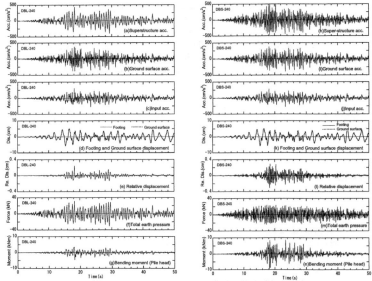

Figure 2: Time histories in non-liquefied soil shaking table tests

than that in case DBL at similar relative displacement.

The total earth pressure during an earthquake may depend not only on the relative displacement but also on the soil inertia around the embedded footing. To investigate the effects of the soil inertia on the total earth pressure, the relations between the lateral seismic coefficient of the surface layer and the total earth pressure are shown in Fig. 4. The lateral seismic coefficient, k_h is defined by

$$k_h = -a_s / g \qquad (2)$$

in which a_s is the average of the accelerations measured at the ground surface and at G.L –0.5m and g is the gravitational acceleration. The data fallen in the first and third quadrants show that the total earth pressure tends to be in phase with the soil inertia, while those in the second and fourth quadrants show that the total earth pressure tends to be out of phase by 180 degrees with the soil inertia. Each black circle in the figure corresponds to the data when the total earth pressure takes its peak in each test. The data from case DBL fallen in all the four quadrants, suggesting that the phase between the two fluctuates with time. The maximum total earth pressure in this case falls in the first quadrant, being in phase with the lateral seismic coefficient. The most data from case DBS fallen in the second and fourth quadrants, suggesting that the total earth pressure tends to be out of phase by 180 degrees with the lateral seismic coefficient. From the above, the phase between the lateral seismic coefficient and the total earth pressure in cases DBL and DBS differ from each other.

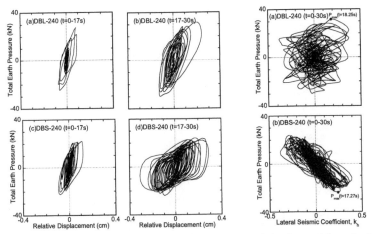

Figure 3: Relation between relative displacement and total earth pressure

Figure 4: Relation between lateral seismic coefficient and total earth pressure

Effects of Soil Inertia on Total Earth Pressure

Effects of soil inertia around the embedded footing on total earth pressure will be investigated, using the earth pressure theory based on the concept "intermediate soil wedge" (Zhang et al, 1998). The equations derived from this theory can evaluate seismic earth pressure against retaining walls under any conditions between the active and passive states and become equal to the Mononobe-Okabe equations at two extreme states, i.e. active and passive states.

Figure 5 schematically shows the changes of soil wedges and earth pressure on both sides of the footing during shaking. The soil wedges on both sides can differ from each other due to relative displacement between footing and soil in such a way that the size of the wedge on the passive side becomes larger than that on the active side. In addition, each of the wedges can further change its dimension and weight depending on the direction of lateral seismic coefficient k_h. The direction of k_h is defined as shown in the figure, in which W is the weight of either of the passive or active soil wedge and $k_h W$ is soil inertia. The plus sign of the lateral seismic coefficient indicates that the soil inertia is in phase with the total earth pressure defined as a difference between passive and active side earth pressures. When $k_h > 0$, the passive side wedge increases its size with decreasing size of the active side wedge as shown in Fig. 5(a). Therefore, the passive side earth pressure increases and the active side earth pressure decreases as shown in Fig. 5(b) and thus the total earth pressure increases with increasing k_h in the positive direction. When $k_h < 0$, the passive side wedge decreases its size with increasing size of the active side wedge as shown in Fig. 5(b). Therefore, the passive side earth pressure deceases and the active side earth

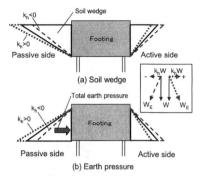

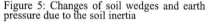

Figure 5: Changes of soil wedges and earth pressure due to the soil inertia

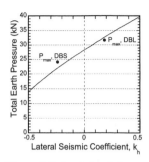

Figure 6: Estimated and observed relation between total earth pressure and k_h

pressure increases as shown in Fig. 5(b) and thus the total earth pressure decreases with increasing k_h in the negative direction.

To evaluate the effects of the soil inertia, the relation between the lateral seismic coefficient and the ultimate total earth pressure defined as a difference between passive and active earth pressures for the same k_h is estimated as shown in Fig. 6. It is assumed that the unit weight of the soil is 18 kN/m^3, the internal friction angle, ϕ is 35 degrees and the wall friction angle is a half of the internal friction angle. Black circles in the figure show the maximum total earth pressure in cases DBL and DBS, respectively. The observed data seem to be consistent with the estimated values. The good agreement suggests that the proposed method gives a reasonable explanation of difference in total earth pressure between the two tests. The total earth pressure is estimated to be 33 kN at k_h=0.2 and 23 kN at k_h=-0.2, respectively. The former is about 1.5 times as large as the latter, indicating the total earth pressure varies significantly depending on its phase relative to the soil inertia around the embedded footing as well as the relative displacement between soil and footing.

Effects of Total Earth Pressure on Bending Moment at Pile Head

To investigate the effects of the total earth pressure on the bending moment at the pile head, the phase between the total earth pressure and the superstructure inertia in cases DBL and DBS are shown in Fig. 7. The data fallen in the first and third quadrants show that the total earth pressure tends to be in phase with the superstructure inertia, while those in the second and fourth quadrants show that the total earth pressure tends to be out of phase by 180 degrees with the soil inertia. A gray line in the figure shows that the soil displacement ΔS is smaller than the footing displacement ΔB, while a black line shows that ΔS is larger than ΔB.

Most of the data from both tests fall in the second and fourth quadrants. This indicates that the total earth pressure is out of phase by 180 degrees with the superstruc-

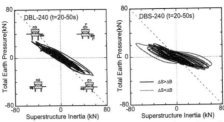

Figure 7: Relation between superstructure inertia and total earth pressure

ture inertia and induces smaller shear force and bending moment at the pile head. In case DBL, a black line is dominant, indicating that ΔS tends to be larger than ΔB when the total earth pressure reaches its peak. In case DBS, on the contrary, ΔS tends to be smaller than ΔB, suggesting that the phase between the superstructure inertia and the total earth pressure in case DBL is different from that in case DBS.

A method for estimating the phase between the superstructure inertia and the total earth pressure proposed by Tamura et al. (2002b) will be modified to take into account the effects of the soil inertia on the total earth pressure. In the method, the soil inertia is assumed to be in phase with the soil displacement, because the acceleration and displacement of a sinusoidal wave have the opposite sign with each other and the acceleration and lateral seismic coefficient also have the opposite sign. Based on this assumption, the total earth pressure is in phase with the soil inertia when the total earth pressure is in phase with the soil displacement. In this case, the total earth pressure will increase with increasing soil inertia. On the other hand, the total earth pressure is out of phase by 180 degrees with the soil inertia when the total earth pressure is out of phase with the soil displacement. In this case, the total earth pressure will decrease with increasing soil inertia. Taking into account the effects of soil inertia on total earth pressure, the mechanism of the total earth pressure can be classified into four types as shown in Fig. 8, based on the natural period of the superstructure under fixed footing condition T_b, the predominant period of the ground T_g, the soil displacement ΔS, and the footing displacement ΔB.

a) If $T_b < T_g$ and $\Delta S > \Delta B$, the total earth pressure tends to be in phase with the superstructure inertia (Fig. 8 (a)). Thus, the bending moment at the pile heads increases due to the total earth pressure. Considering that the soil displacement is in phase with the total earth pressure as shown in Fig. 8(a), the total earth pressure, P_+ will increase with increasing soil inertia.

b) If $T_b > T_g$ and $\Delta S > \Delta B$, the total earth pressure tends to be out of phase by 180 degrees with the superstructure inertia (Fig. 8 (b)). Thus, the bending moment at the pile heads decreases due to the total earth pressure. Considering that the soil displacement is in phase with the total earth pressure as shown in Fig. 8(b), the total earth pressure, P_+ will increase with increasing soil inertia.

c) If $T_b < T_g$ and $\Delta S < \Delta B$, the total earth pressure tends to be out of phase by 180

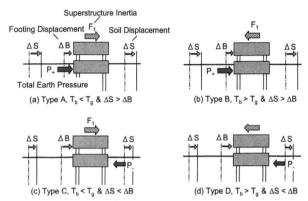

Figure 8: Phase difference between total earth pressure and superstructure inertia

degrees with the superstructure inertia (Fig. 8 (c)). Thus, the bending moment at the pile heads decreases due to the total earth pressure. Considering that the soil displacement is out of phase by 180 degrees with the total earth pressure as shown in Fig. 8(c), the total earth pressure, P. will decrease with increasing soil inertia.

d) If $T_b > T_g$ and $\Delta S < \Delta B$, the total earth pressure tends to be in phase with the superstructure inertia (Fig. 8 (d)). Thus, the bending moment at the pile heads increases due to the total earth pressure. Considering that the soil displacement is out of phase by 180 degrees with the total earth pressure as shown in Fig. 8(d), the total earth pressure, P. will decrease with increasing soil inertia.

The proposed method will be applied to the test results in cases DBL and DBS. In case DBL, ΔS tends to be larger than ΔB and T_b is longer than T_g. These conditions correspond to Type B in Fig. 8. The total earth pressure in this case increases with increasing soil inertia and tends to be out of phase by 180 degrees with the superstructure inertia. In case DBS, on the contrary, ΔS tends to be smaller than ΔB and T_b is shorter than T_g. These conditions correspond to Type C. The total earth pressure in this case decreases with increasing soil inertia and tends to be out of phase by 180 degrees with the superstructure inertia. The important point to note is that the phase difference between the superstructure inertia and the total earth pressure in case DBL is different from that in case DBS, though the superstructure inertia and the total earth pressure tend to be out of phase in both models.

LIQUEFACTION SHAKING TABLE TESTS

Dynamic Response of Soil-Pile-Superstructure System

Figure 9 shows time histories of acceleration of the superstructure, ground surface and input motion, footing and ground surface displacement, relative displacement between footing and ground surface, total earth pressure and bending moment at the pile head in cases SBL and SBS, after the study by Tamura et al. (2002a). The predominant period of the ground surface acceleration and the total earth pressure become longer and the amplitude of the ground surface displacement and the bending moment increase rapidly at 8 seconds in both tests.

Figure 10 shows the vertical distributions of the excess pore water pressure. The excess pore water pressure begins to increase at 8 seconds in both models, suggesting that the changes of the ground surface acceleration, the total earth pressure, the

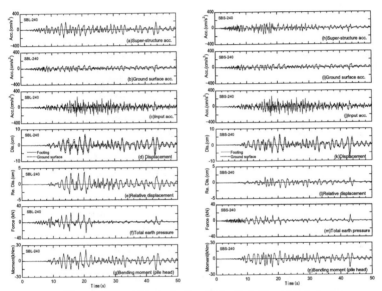

Figure 9: Time histories in liquefied soil shaking table tests

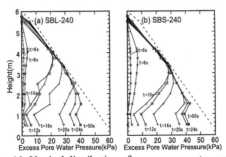

Figure 10: Vertical distribution of excess pore water pressure

ground surface displacement and the bending moment were caused by the soil liquefaction. Soil liquefaction developed from the upper layer to the bottom. The lower part of the saturated sand layer almost liquefied at about 20 seconds.

All dynamic responses in the liquefaction tests were quite different from those in the non-liquefaction tests in spite of the same input motion. It is interesting to note that the bending moment in the liquefaction tests is much larger than that in the non-liquefaction tests, though the superstructure inertia in the liquefaction tests is smaller than that in the non-liquefaction tests. Considering that the bending moment depends not only on the superstructure inertia but also on the total earth pressure, the total earth pressure plays an important role in controlling bending moment in piles.

Effects of Soil Inertia and Relative Displacement on Total Earth Pressure

To investigate the mechanism of the total earth pressure, the relations between the lateral seismic coefficient of the surface layer and the total earth pressure for two time segments (0-8 and 8-50 sec.) are shown in Fig. 11. Black circles in the figure correspond to the data when the total earth pressure takes its peak in each test. The data from case SBL fall in all the quadrants until 8 seconds, indicating the phase between the lateral seismic coefficient and the total earth pressure fluctuates with time. The data from case SBS fall in the second and fourth quadrants until 8 seconds, indicating the total earth pressure tends to be out of phase by 180 degrees with the lateral seismic coefficient. The phases in cases SBL and SBS before liquefaction are similar to those in cases DBL and DBS, respectively. After 8 seconds, the data from both models fall in the first and third quadrants, indicating the total earth pressure tends to be in phase with the lateral seismic coefficient. The lateral seismic coefficients at the maximum total earth pressure are smaller than 0.1 in both models, indicating the effect of the soil inertia on the total earth pressure is relatively small.

Figure 12 shows the relations between the total earth pressure and the relative displacement in cases SBL and SBS. The relations between two are elliptical in shape in both models until 8 seconds. It becomes nonlinear with the development of the pore water pressure from 8 to 20 seconds. The total earth pressure gets small with cyclic loading from 20 to 50 seconds. The relative displacement in case SBL tends to be larger than that in case SBS after 8 seconds. The total earth pressure in case SBL also tends to be larger than that in case SBS. Taking into account the little effect of the

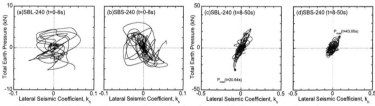

Figure 11: Relation between lateral seismic coefficient and total earth pressure

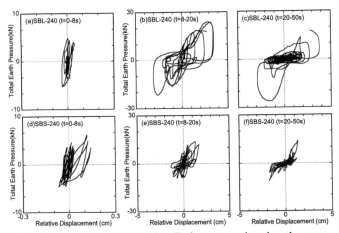

Figure 12: Relation between relative displacement and total earth pressure

soil inertia on the total earth pressure, the total earth pressure is affected mainly by the relative displacement.

Effects of Total Earth Pressure on Bending Moment at Pile Head

Figure 13 shows the relations between the total earth pressure and the superstructure inertia in the two tests (Tamura et al. 2002b). The total earth pressure tends to be out of phase by 180 degrees with the superstructure inertia in both models until 8 seconds. The phase in cases SBL and SBS is similar to those in cases DBL and DBS as shown in Fig. 7. The total earth pressure, which is equivalent to about 60-70 percent of the superstructure inertia, counteracts the inertial force transmitted from the superstructure to the pile heads in both models. Therefore, the bending moment is very small until 8 seconds as shown in Fig. 9.

The relations between the two in Fig. 13 in both tests changed after 8 seconds, significantly. The total earth pressure tends to be in phase with the superstructure inertia. ΔS tends to be larger than ΔB when the total earth pressure reaches its peak. In addition, T_b is shorter than T_g due to the liquefaction in both models. These conditions correspond to Type A in Fig. 8. The total earth pressure in this case tends to be in phase with the superstructure inertia. Therefore, the bending moments increase rapidly after 8 seconds as shown in Fig. 9.

In general, displacement and a predominant period of liquefied soil become very large and long. This suggests that the phase between the superstructure inertia and the total earth pressure corresponds to Type A in Fig. 8 in most cases during liquefaction except for very tall buildings. Therefore, it seems reasonable to assume that any non-liquefied crust on a liquefied soil layer imposes severe conditions on piles.

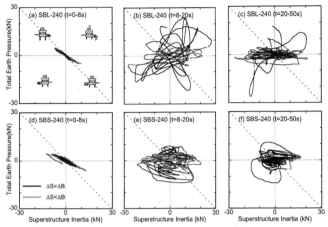

Figure 13: Relation between superstructure inertia and total earth pressure

CONCLUSION

This paper examines earth pressure acting on an embedded footing and its effects on pile forces, based on both liquefaction and non-liquefaction tests using a large-scale laminar shear box. The following conclusions are drawn:

(1) The total earth pressure in the non-liquefaction tests varies significantly depending on its phase relative to the soil inertia around the embedded footing as well as on the relative displacement between soil and footing.

(2) The total earth pressure in the liquefaction test, by contrast, depends mainly on the relative displacement because the soil inertia gets small in liquefied soil.

(3) The total earth pressure in the non-liquefaction tests tends to be out of phase by 180 degrees with the superstructure inertia, reducing the shear force and bending moment at the pile head.

(4) The total earth pressure in the liquefaction tests tends to be in phase with the superstructure inertia, making the bending moment at the pile head large. A method for estimating the total earth pressure considering its phase relative to the superstructure inertia as well as the effects of soil inertia has been proposed. The proposed method gives a reasonable explanation of the difference in earth pressure between different tests.

ACKNOWLEDGMENTS

The non-liquefaction shaking table tests were made under Special Project for Earthquake Disaster Mitigation in Urban Areas, supported by Ministry of Education, Culture, Sports, Science and Technology (MEXT). The liquefaction tests were jointly conducted by NIED, Tokyo Institute of Technology, Kajima Corp., Taisei Corp., Takenaka Corp., Nippon Steel Corp. and Tokyo Soil Research Corp. The contribution and support of the above organizations are appreciated.

REFERENCES

Fujii, S., Isemoto, N., Satou, Y., Kaneko, O., Funahara, H., Arai, T. and Tokimatsu, K. (1998). "Investigation and analysis of a pile foundation damaged by liquefaction during the 1995 Hyogoken-Nambu earthquake", Soils and Foundations, Special Issue on Geotechnical Aspects of the 1995 Hyogoken Nambu Earthquake, No. 2, pp. 179-192.

Horikoshi, K., Tateishi, A., and Fujiwara, T. (1998). "Centrifuge Modeling of a Single Pile Subjected to Liquefaction-induced Lateral Spreading", Soils and Foundations, Special Issue on Geotechnical Aspects of the 1995 Hyogoken Nambu Earthquake, No. 2, pp. 193-208.

Miyamoto,Y., Sako Y, Koyamada K. and Miura K.(1997). "Response of pile foundation in liquefied soil deposit during the Hyogo-ken Nanbu Earthquake of 1995", Journal of Struct. Constr. Engng. AIJ, No.493 , pp.23-30. (in Japanese)

Murono, Y. and Nishimura, A.(2000), "Evaluation of Seismic Force of Pile Foundation Induced by Inertial and Kinematic Interaction", Proc., of 12th World Conf. on Earthq. Engrg., Reference No. 1496.

Oh-Oka, H., Fukui, M., Hatanaka, M., Ohara, J. and Honda, S. (1998). "Permanent deformation of steel pipe piles penetrating compacted fill at wharf on port Island", Soils and Foundations, Special Issue on Geotechnical Aspects of the 1995 Hyogoken Nambu Earthquake, No. 2, pp. 147-162.

Sato, M., Shamoto, Y. and Zhang, J. -M.(1995). "Soil Pile-structure during Liquefaction on Centrifuge", Proc., 3rd International Conference on Recent Advances in Geotechnical Earthquake Engineering and Soil Dynamics, 1, pp. 135-142.

Tamura, S., Tsuchiya, T., Suzuki, Y., Fujii, S., Saeki, E. and Tokimatsu, K.(2000). "Shaking Table Tests of Pile Foundation on Liquefied Soil Using Large-scale Laminar Box (Part 1: Outline of Test)", 35th Japan National Conference on Geotechnical Engineering, pp.1907-1908. (in Japanese)

Tamura, S., Tokimatsu, K., Miyazaki, M., Yahata, K. and Tsuchiya, T. (2002a), "Seismic earth pressure acting on embedded footing based on liquefaction test using large scale shear ", Journal of Struct. Constr. Engng., No.554 , pp.95-100. (in Japanese)

Tamura, S., Tokimatsu, K., Uchida, A., Funahara, H. and Abe, A.(2002b). "Relation between Seismic Eearth Pressure Acting on Embedded Footing and Inertial Force Based on Liquefaction Test Using Large Scale Shear Box", Journal of Struct. Constr. Engng., No.559 , pp.129-134. (in Japanese)

Tokimatsu, K. and Asaka, Y. (1998). "Effects of liquefaction-induced ground displacements on pile performance in the 1995 Hyogoken-Nambu Earthquake", Soils and Foundations, Special Issue of Geotechnical Aspects on the 1995 Hyogoken Nambu Earthquake, No. 2, pp. 163-177.

Tokimatsu, K., Suzuki, H., Suzuki, Y. and Fujii, S.(2002). "Evaluation of Lateral

Subgrade Reaction of Pile during Soil Liquefaction Based on Large Shaking Table Tests", Journal of Struct. Constr. Engng. , No.553 , pp.57-64. (in Japanese)

Tokimatsu, K., Suzuki, H. and Sato, M.(2004). "Influence of inertial and kinematic components on pile response during earthquake", Proc. of 11th International Conference on Soil Dynamics and Earthquake Engineering & 3rd International Conference on Earthquake Geotechnical Engineering, Berkeley, Vol.1, pp. 768-775.

Wilson, D. W., Boulanger, R. W. and Kutter, B. L.(2000). "Observed seismic lateral resistance of liquefying sand", J. Geotechnical and Geoenvironmental Engineering, ASCE, 12(10), pp.898-906.

FIELD INVESTIGATION AND ANALYSIS STUDY OF DAMAGED PILE FOUNDATION DURING THE 2003 TOKACHI-OKI EARTHQUAKE

Kohji Koyamada[1], Yuji Miyamoto[1] and Kohji Tokimatsu, Member, ASCE[2]

ABSTRACT

Konan junior high school in Hokkaido, the northernmost main island of Japan, suffered severe damage to its superstructure and pile foundations during the Tokachi-oki Earthquake (M=8.0), on September 26, 2003. The objective of this study is to verify the main factors causing this damage by field investigations and earthquake response analyses. Detailed field surveys were carried out on the superstructure and pile foundations immediately after the quake. Analyses of the structure of the pile foundation were conducted using a numerical model taking into account the effect of soil nonlinearity. Strong motions observed near the school were employed as the input motion to the analysis model. It is found that compression failures at the pile heads were caused not only by the inertial force from its superstructure but also by large displacement of a very soft clay, leading to differential settlement as well as shear cracks in walls of the superstructure.

INTRODUCTION

The Tokachi-Oki Earthquake, which occurred on September 26, 2003, affected the entire area of Hokkaido, Japan, causing extensive damage. The resulting damage has been described in the results of several investigations (e.g. Architectural Institute of Japan, 2004, etc.), which showed that there were far fewer fatalities and less heavy damage than would have been expected for a magnitude 8.0 earthquake. However, ground failures including soil liquefaction occurred, resulting in settlement and tilting of many wooden houses located up to about 270km from the epicenter, e.g. Sapporo-city. Konan junior high school in Atsuma-town, southeast of Sapporo-city,

[1] Kobori Research Complex, Kajima Corp., Tokyo, Japan
[2] Tokyo Institute of Technology, Tokyo, Japan

was severely damaged and a large number of shear cracks occurred on the walls of school building due to differential settlement. It was shown from a detailed field survey (Kajima Corp., 2003) that this damage was caused by the collapse of the pile foundation.

This paper firstly describes the results of the investigation for the superstructure and pile foundations. Next, the pile damage is verified by earthquake response analyses using observed ground motions. Finally, the damaged pile is extracted and the true state of the damage to the pile is confirmed, and the accuracy of the analysis results is examined.

OUTLINES OF THE EARTHQUAKE AND THE DAMAGED BUILDING

Fig. 1 shows an epicenter of the 2003 Tokachi-oki Earthquake and the location of the damaged building, Konan junior high school. The earthquake occurred with its epicenter offshore of Tokachi, about 206km southeast from the site. National Research Institute for Earth Science and Disaster Prevention (NIED) has deployed a lot of downhole strong motion stations, called KiK-net system, including one about 50m from the damage building. This enables one to obtain strong motion accelerograms recorded at the ground surface and a depth of 153m during the earthquake. The peak ground accelerations in the EW direction were 376cm/s^2 at the ground surface and 53cm/s^2 in the downhole.

Fig. 2 shows a plan of the school building, which was constructed in 1980. The superstructure of the school building comprises a three-story RC structure. Fig. 3 shows a bore log and a pile foundation. It is a pre-stressed high-strength concrete (PHC) pile with a diameter of 40cm and a length of 28.5m. The surface soil above 30m depth comprises of very soft layers of peat, clay and sandy silt, which is underlain by a layer of dense gravel with Vs=320m/s. A soil investigation shows that the site comprises of almost horizontally stratified soil deposits.

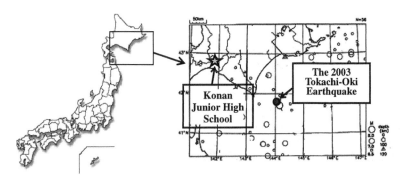

Figure 1 : Epicenter of the 2003 Tokachi-oki Earthquake and location of Konan junior high school

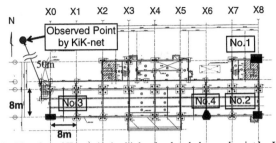

Figure 2 : Plot plan of Konan junior high school and observed point by KiK-net
(Numbers indicate portions of investigated piles)

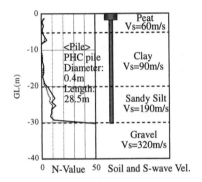

Figure 3 : Boring log and pile foundation

FIELD INVESTIGATION OF DAMAGE TO SUPERSTRUCTURE AND PILE FOUNDATION

Field investigations on the superstructure and pile foundations were performed immediately after the quake.   The main findings from these surveys are summarized below.

Field Surveys of Superstructure

Fig. 4 shows the investigation results for the superstructure.   The maximum horizontal displacement of the roof was 56mm and the maximum tilt angle of the column was 1/220 at the northern east part.   The maximum settlement of the first floor was 110mm at the east end and the maximum slope angle of the floor was 1/160 between X6 and X7.

Fig. 5 shows shear cracks in the wall and Photo 1 shows a close-up of the area where the maximum cracks occurred.   The shear cracks were concentrated between X6 and X7 due to the differential floor settlements.   The maximum crack width was

6mm and the maximum crack length was about 1m.   On the basis of these results, the decision for the reconstruction of the building was taken.

Excavation Surveys Around the Pile Head

Excavation surveys were conducted around the four perimeter piles, labeled as Nos. 1-4 in Fig. 2.   Photo 2 shows the excavation and damage to the northeast pile head (No.1 in Fig. 2).   Not only pile No.1 but also all the other investigated perimeter

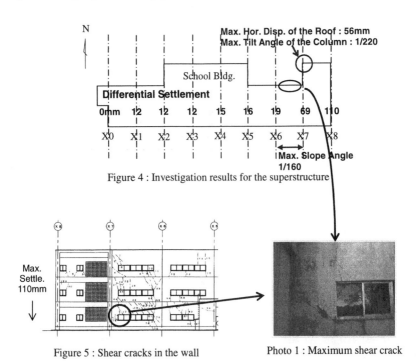

Figure 4 : Investigation results for the superstructure

Figure 5 : Shear cracks in the wall

Photo 1 : Maximum shear crack

(a) Excavation survey   (b) Damage to pile head

Photo 2 : Excavation survey around the pile head (No.1 in Fig. 2) piles were damaged by compression failure with flexural cracks at the pile head. The main finding from these surveys is that compression failures of the pile head induced differential settlements of the superstructure, leading to the necessity for demolition.

SIMULATION ANALYSIS FOR DAMAGE TO PILE FOUNDATION

Simulation analyses were performed in order to verify the causes of the damage to the pile foundations. Evaluation of soil deformation during an earthquake is very important in examining pile stresses. As mentioned earlier, ground motions during the quake were obtained by NIED at two elevations: the ground surface and downhole at a depth of 153m. In this study, not only the response analysis of a one-dimensional soil column but also that of a soil-pile-structure system is conducted, using the downhole recorded motion as an input. The objectives of these analyses are to estimate ground response and then to investigate the major cause of the pile damage.

Analyses Model for Soil-Pile-Structure System

Ground motions were calculated by means of a nonlinear response analysis method, taking into account the strain dependency of the soil dynamic property (Fujimura and Okimi, 2000). Table 1 shows the soil constants and Fig. 6 shows the relations of the shear modulus ratios and the damping factors with the shear strains for the soil deposits. The S-wave velocities were based on soil survey results at the site by NIED. The soil nonlinearities shown in Fig. 6 were set on the basis of laboratory test results from a large number of natural soil samples in Japan (Koyamada et. al., 2003), taking into account of confining stresses and plasticity. Since neither ground subsidence nor sand boils were observed at the site, effects of liquefaction were not considered.

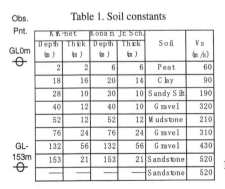

Table 1. Soil constants

	K K-net		Konan Jr. Sch.		Soil	Vs
	Depth (m)	Thick (m)	Depth (m)	Thick (m)		(m /s)
	2	2	6	6	Peat	60
	18	16	20	14	Clay	90
	28	10	30	10	Sandy Silt	190
	40	12	40	10	Gravel	320
	52	12	52	12	Mudstone	210
	76	24	76	24	Gravel	310
	132	56	132	56	Gravel	430
	153	21	153	21	Sandstone	520
	—	—	—	—	Sandstone	520

Obs. Pnt. GL0m, GL-153m

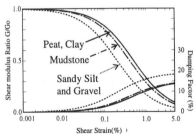

Figure 6 : Relation of shear modulus and damping factor with shear strain

Seismic response analyses of the structure supported on the pile foundations were conducted using a beam-interaction spring model (Miyamoto et. al., 1995), as shown in Fig. 7. The superstructure and pile foundations were idealized by a one-stick model of lumped masses and beam elements.   The lumped masses of the pile foundations were connected to the free field soil through nonlinear lateral and shear interaction springs.   A linear rotational spring related to the axial stiffness of the piles was also incorporated at the pile head.   The initial values of the lateral and shear interaction soil springs of the pile groups were obtained using Green's functions by ring loads in a layered stratum (Kausel and Peek, 1982).   The nonlinear soil spring was a function of the relative displacement between soil and pile.

Table 2 shows analysis parameters for the superstructure.   The relationship between the shear force and the shear deformation of the superstructure is assumed to be trilinear.   The nonlinear properties of the pile foundations are incorporated into the relationships between the bending moments and the curvature, which are evaluated by a static pushover analysis with a fiber-model, and assumed to be also trilinear (Concrete Pile Installation Technology Association), as shown in Fig. 8.

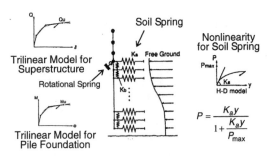

Figure 7 : Numerical model of soil-pile-structure system

Table 2. Analyses parameter for superstructure

F1	Height (cm)	Weight (kN)	Initial Stiffness K1 (kN/cm)	Shear Force at 1st Reflect Pnt Q1 (kN)	2nd Stiffness K2 (kN/cm)	Ultimate Resistance Force Qu (kN)
3	365	7110	15300	5150	4580	10300
2	370	9140	14030	4720	4130	9440
1	375	11090	24200	8130	6990	16270

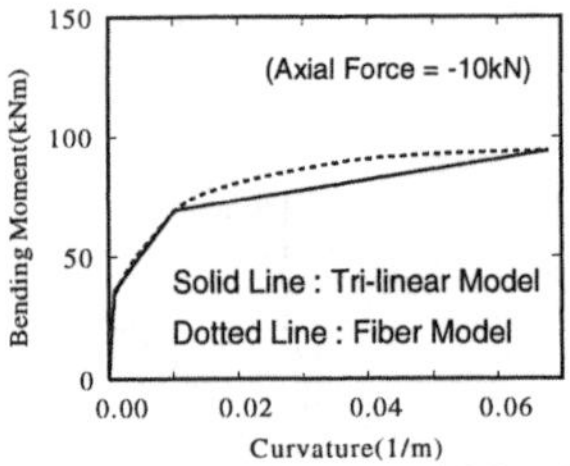

Figure 8 : Trilinear model for pile foundation

Simulation Analyses Results for Ground Motion

Fig. 9 shows the acceleration time histories of the observed and computed ground surface motion together with the observed downhole motion in addition to their acceleration response spectra for a damping ratio of 5%.   The response spectrum of the observed ground motion has a peak at a period of about 2s, which is in good agreement with that of the computed ground surface motion.

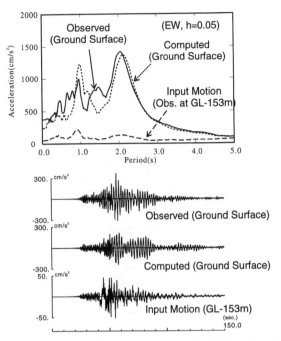

Figure 9 : Acceleration response spectra for a damping ratio of 5% and acceleration time histories of ground motions

Fig. 10 shows the acceleration response spectra for a damping ratio of 5% at several depths including the ground surface and depths of 20, 76 and 153m. The spectral peak occurring at the period of 2s increases by ten times between a depth of 153m and the ground surface. In particular, it is amplified by about 7 times between a depth of 76m and the ground surface. This indicates that the predominant motions at the period of 2s are mainly amplified in the layer above 76m depths.

Fig. 11 shows the distributions of the maximum relative displacements with respect to the pile tip, the maximum shear strains and shear modulus ratios in the layers above 30m depths. The soil displacements are amplified in the clay layer and exceed 20cm at the ground surface. The shear strains become large at depths of 6m and 20m by about 2% and 3%, respectively. Soil stiffness is reduced corresponding to the shear strain and the minimum shear modulus ratio is about 0.2 at the depth of 20m.

Dynamic Response of the Superstructure

Fig. 12 shows the acceleration response spectra for a damping ratio of 5% on the roof and the first floor of the building, compared with that of the ground surface.

The response spectra on the roof are significantly amplified near a period of 0.35s, which corresponds to the fundamental natural period of the building.

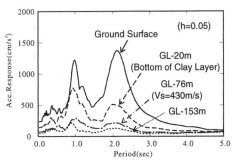

Figure 10 : Acceleration response spectra for a damping ratio of 5 % at the ground surface and depths of 20m, 76m and 153m

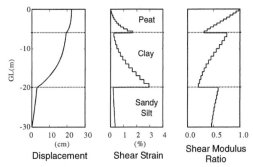

Figure 11 : Distributions of the maximum relative displacements with respect to the pile tip, the maximum shear strains and shear modulus ratios in the layers above 30m depths

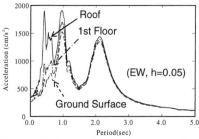

Figure 12 : Acceleration response spectra for a damping ratio of 5% of the superstructure

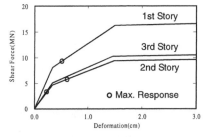

Figure 13 : Maximum superstructure responses for the trilinear model

Fig. 13 shows the maximum superstructure responses for the tri-linear model.   The maximum shear deformation responses of the first and second stories are 1/682 and 1/617, respectively, and the maximum shear force responses are smaller than the ultimate resistance forces.   This suggests that many shear cracks in the structure wall were not caused by the shear force but by the differential settlement of the superstructure.

Verification for the Pile Damage

Fig. 14 compares the distributions of maximum bending moments and shear forces (total component) with those induced by the inertial force of the superstructure (inertial component) and those by the ground displacement (kinematic component). The kinematic components were computed using the analysis model without the superstructure and the inertial components were obtained by subtracting the kinematic components from the total components in the time domain. The bending moment at the pile head, which consists of both inertial and kinematic components, exceeds the ultimate bending moment M_u (130kN*m).   In addition, the bending moment at a depth of 20m, which consists mainly of kinematic component, also exceeds Mu.

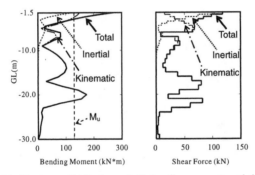

Figure 14 : Maximum distributions of pile bending moments and shear forces

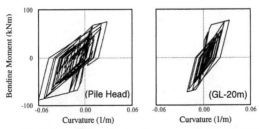

Figure 15 : Relationships of the bending moments with the curvature
at the pile head and at the depth of 20m

In order to evaluate the degree of damage to the pile, analyses incorporating pile nonlinearity were conducted.   Fig. 15 shows relationships of the bending moments with the curvature at the pile head and at the depth of 20m. The maximum curvature at the pile head was about 0.06/meter and is consistent with the damage obtained from the excavation surveys.   The curvature at the depth of 20m also reached about 0.03/meter, indicating that bending cracks might have occurred at this depth.

From these analyses results, it is shown that damage to the pile occurred not only at the pile head but also at a depth of 20m.

EXTRACTION SURVEY OF PILE FOUNDATION

In order to confirm the true state of the damage and to verify the accuracy of the analysis results, an extraction survey of the pile foundation was conducted.   Photo 3 shows the process of extracting the damaged pile (No.2 in Fig. 2) and Photo 4 shows a full view of the extracted pile.   Although about twenty-five years had passed since the building's construction, there was no rust on the steel tip of the pile.

Photo 3 : Process of extracting the damaged pile (No.2 in Fig. 2)

Photo 4 : Full view of the extracted pile

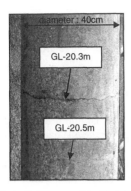

Photo 5 : Discovered cracks of pile near the depth of 20m

Photo 6 : Investigation of the inner side for the PHC pile by cutting out the area of the largest crack

From the investigation, a large crack was discovered at about 20m in depth, as shown in Photo 5. Bending cracks occurred at depths of 20.0m, 20.3m and 20.5m. The crack at 20.3m was the largest and its maximum width was about 1.0mm. Moreover, the situation of the inner side for the PHC pile was checked by cutting out the area of the largest crack, as show in Photo 6. A crack occurred around the pile on the inner surface as well as the outer surface, and its maximum width was about 0.5mm. The other cracks, at 20.0m and 20.5m, were small and both were 0.1mm in width.

CONCLUSIONS

The 2003 Tokachi-Oki Earthquake caused severe damage to Konan junior high school in Japan. Well-documented field investigations of the superstructure and pile foundations and earthquake response analyses and taking into account the soil nonlinearity lead to the following conclusions:

(1) The maximum horizontal displacement and tilt angle of the roof were 56mm and 1/220, and the maximum settlement and slope angle of the first floor were 110mm and 1/160 between two adjacent columns. The shear cracks on the walls were caused by the differential settlements of the floor.

(2) Excavation surveys clarified that all the investigated perimeter piles were damaged by compression failure with flexural cracks at the pile head. These pile failures induced the differential settlements of the superstructure, leading to the decision of demolishing the school.

(3) The observed motion at the ground surface had a large component at a period of about 2s that was mainly amplified in the layer above 76m depths. The soil displacements were amplified in the upper clay layer above 20m in depth and soil stiffness decreased at the depth of 20m. The computed ground surface motion showed a good agreement with the observed motion.

(4) The maximum shear deformations of the first and second stories were very small: 1/682 and 1/617, respectively, suggesting that many large cracks in the

superstructure wall were mainly due to the differential settlements and not due to the seismic motions.

(5) Pile response analyses indicated that the maximum curvature at the pile head was about 0.06/meter and is consistent with the damage obtained from the excavation surveys.   The curvature at the depth of 20m also reached about 0.03/meter, indicating that bending cracks might have occurred at this depth.

(6) The extraction survey of the damaged pile revealed that large cracks occurred not only on the outer but also on the inner surface of the pile near a depth of 20m. This verified the accuracy of the simulation analyses for the pile damage.

This study suggests that a pile foundation could be damaged by soil displacement during a large earthquake if the site has a very soft layer, even in a horizontally stratified deposit.   It is thus crucial to consider soil nonlinearity for seismic design of pile foundations.

ACKNOWLEDGMENTS

We are grateful to the staffs of the Atsuma municipal government for their assistance throughout this study.   Special thanks go to Mr. S. Nishio and Mr. M. Morimoto. Without their help, this work would not have been possible.   We would also like to thank Mr. H. Yokoi, Nishimura Design Office, and Mr. H. Watanabe, Asahi Kasei Construction Materials, for providing information concerning the pile performance.   Our gratitude is extended to Mr. T. Fukuda, Mr. H. Tomizawa and Mr. N. Arie, Kajima Corp., for their implementation of the full field investigations.

REFERENCES

Architectural Institute of Japan, Hokkaido Bureau (2004) "Report on the Damage Investigation of the Tokachi-oki Earthquake in 2003", 143pp. (in Japanese).
Fujimura, K., and Okimi, Y. (2000). "The response of three-dimensional multiple shear model for liquefaction analysis" *Proc., 35th Japan National Conf. on Geo. Eng.*, The Japanese Geotechnical Society, pp. 955-956. (in Japanese)
Kajima Corp. (2003). "Report on the Damage to Konan junior high school in Atsuma-town during the 2003 Tokachi-oki Earthquake", 120pp. (in Japanese)
Kausel E., and Peek R. (1982) "Dynamic Loads in the Interior of a Layered Stratum-An Explicit Solution", *Bulletin of the Seismological Society of America*, Vol. 72, pp. 1459-1481.
Koyamada, K., Miyamoto, Y., and Miura, K. (2003). "Nonlinear Property for Surface Strata from Natural Soil Samples" *Proc., 38th Japan National Conf. on Geo. Eng.*, The Japanese Geotechnical Society, pp. 2077-2078. (in Japanese)
Miyamoto, Y., Sako, Y., Kitamura, E., and Miura, K. (1995) "Earthquake Response of Pile Foundation in Nonlinear Liquefiable Sand Deposit", *Journal of Structural and Construction Engineering*, AIJ, No. 471, pp.41-50 (in Japanese)
National Research Institute for Earth Science and Disaster Prevention. KiK-net HP, http://www.kik.bosai.go.jp

PUSH-OVER ANALYSES OF PILES IN LATERALLY SPREADING SOIL

Scott A. Ashford, Member, ASCE[1], Teerawut Juirnarongrit[2]

ABSTRACT

This paper presents the assessment of push-over analyses of single pile and pile groups in laterally spreading ground using p-y approach. The results from full-scale blast induced lateral spreading experiments in Japan were used to evaluate the numerical model. For the single pile, the pile responses were determined by imposing the measured free field displacements to the Winkler spring model. For pile groups, the piles in the group were modeled as an equivalent single pile, together with reduced stiffness p-y springs to account for the effect of pile group. A rotational spring was also introduced into the model to represent the rotational pile group stiffness. The results from the analyses for both single pile and pile groups were in good agreement with the measured responses from the experiments, thus suggesting that this method may be used in common design practice for piles in laterally spreading ground problems.

INTRODUCTION

A pseudo-static pushover analysis using the p-y analysis method is widely used in current design practice to analyze the response of laterally loaded piles due to its simplicity in modeling compared to the 2D or 3D finite element method (FEM). Despite its simplicity, the p-y approach has the capability to provide key design parameters (i.e., pile maximum moment and pile head displacement). Application of this method in current design practice is mainly focused on the analysis of piles under

---

[1] Associate Prof., Department of Structural Engineering, Univ. of California, San Diego, La Jolla, CA 92093-0085.
[2] Postgraduate Researcher, Department of Structural Engineering, Univ. of California, San Diego, La Jolla, CA 92093-0085.

inertial loading, with movement relative to a stationary soil mass (Reese et al. 1974; Matlock 1970). For the case of moving soil mass such as piles in lateral spreading soil, the moving soil mass will exert the load on the pile and displace the pile a certain amount depending on the relative stiffnesses between the pile and the soil. For this kind of application, it requires the free-field soil movement (see Figure 1) as an input to the boundary ends of the Winkler springs (Boulanger et al. 2003; Meyersohn 1994; Tokimatsu and Asaka 1998). When the expected free-field displacement is large enough to cause the ultimate pressure of laterally spreading soils to be fully mobilized, the ultimate pressure, instead of free-field soil displacement, may be used Estimates of the ultimate pressure of liquefied soils have been developed based on calibration with piles damaged in the 1995 Kobe earthquake (JRA 2002), and centrifuge test data (e.g., Dobry et al. 2003).

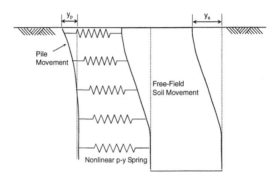

Figure 1:   The p-y analysis model for pile subjected to lateral spreading

Though the method is relatively simple and appears to be one of the most attractive methods for practicing engineers, application of this method to piles in laterally spreading problems is rare in the literature, especially for the case of pile groups due to the limited availability of physical data. As such, the main objective of this paper is to assess the applicability of push-over analysis using the p-y method for single piles and pile groups subjected to lateral spreading problems. The assessment was performed by comparing the results from numerical modeling with the results from the full-scale lateral spreading experiments in the Port of Tokachi, Japan.

SITE DESCRIPTION

Two full-scale lateral spreading experiments were conducted at the Port of Tokachi, Hokkaido Island, Japan to study the behavior of single piles and pile groups subjected to lateral spreading. A layout of the test site for the first test is shown in Figure 2. The test site was approximately 25 m wide by 100 m long with 4% surface slope test bed. The site was bordered by a traditional design of quay walls (e.g., no consideration of seismic force in the design) on one end. The quay wall was fixed to H-piles using a series of tie-rods to reduce the movement of quay wall. The test piles were located 19.0 m away from the quay wall, which consisted of a UCSD single pile, a group of Waseda University (WU) single piles, a 4-pile group, and a 9-pile group. The piles were instrumented with strain gauges to measure moments during lateral spreading. All the single piles had the free-head condition, while the pile groups were fixed to a rigid pile cap. Controlled blasting was used to liquefy the soil at the test site, and thus induce lateral spreading. The blast holes were spaced at 6.0 m on center in the regular grid pattern with the blasting starting from the southwest corner of the embankment and continuing toward the quay wall as shown in Figure 2. More details on the site information and detail for the second lateral spreading test can be obtained elsewhere (Ashford and Juirnarongrit 2004).

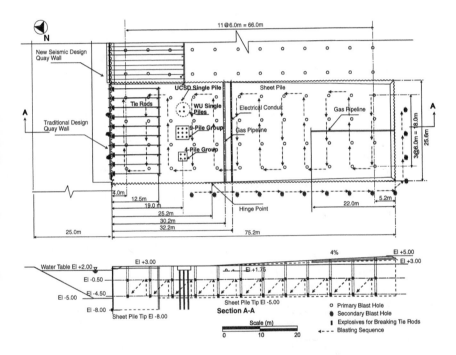

Figure 2: Site layout for 1st lateral spreading test

The soil at the test site consisted of 7.5 m of hydraulic fill, underlain by 1 m of medium dense sand overlying a very dense gravel layer (see Figure 3). The ground water table was approximately 1 m below the ground surface. The hydraulic fill consisted of a 4-m layer of very loose to loose silty sand (SM), underlain by a 3.5-m layer of very soft lean to fat clay with sand (CL to CH). Using the US criterion for liquefaction susceptibility evaluation (Seed and Idriss 1971), the first and second sand layers below the ground water table are susceptible to liquefaction. A summary of soil properties used in the push-over analyses is also given in Figure 3.

PUSH-OVER ANALYSES

Single Pile

To predict the behavior of piles subjected to lateral spreading, the free-field soil displacement, y_s, needs to be known first, then imposed to the boundary ends of the Winkler soil springs along depths as shown in Figure. 3. The free-field soil movement profiles (i.e., no influence from pile foundations) at the end of both the first and second experiments were obtained from the measured soil displacement profiles of a slope inclinometer between the pile groups. Based on these data, simplified linear displacement profiles of the free-field soil movements were used for the boundary condition at the end of soil springs with the largest displacements at the ground surface of 0.43 m for the first test and additional 0.46 m from the second test for a total of 0.89 m. Soil springs at different depths were calculated based on standard p-y springs available in the literature. The p-y curves for non-liquefied cohesionless soil were developed based on Reese et al.'s (1974) recommendations, while the p-y curves for soft clay were obtained based on Matlock's (1970) recommendations. Since the maximum response of the piles due to lateral spreading occurred at the end of the test, where the soil had already been liquefied, the p-y curves for liquefied soil were used for the saturated sand layers. Details of p-y curves for liquefied sand used in this study are described below.

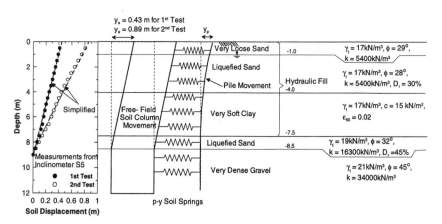

Note: γ_t = unit weight, ϕ = friction angle, c = soil cohesion, k = subgrade reaction constant,
ε_{50} = strain corresponding to one-half the maximum principal stress difference

Figure 3: Soil displacement profiles from 1st and 2nd tests at Tokachi, and p-y analysis model for single pile

Review of current research based on results from various physical modeling techniques suggests that the characteristics of p-y curves of liquefied soil may depend on the initial relative density of the soil.   For loose sand with relative density, D_r, of less than 40% (i.e., Wilson et al. 2000; Tokimatsu et al. 2001), the p-y curves are flat, inferring that the soil pressure from liquefied soil is negligible, while for medium dense and dense sand with relative density of greater than 55% (i.e., Ashford and Rollins 2002; Wilson et al. 2000; Tokimatsu et al. 2001) the p-y curves are concave up due to soil dilation.   The relative densities at the Tokachi test site was slightly over 30% for the first 4-m of the sand layer and about 45% for the second sand layer at depths between 7.5 m and 8.5 m (see Figure 3).   Based on the low initial relative density, the liquefied soil layer at the Tokachi site should not provide any resistance to pile movement. Therefore, zero soil spring stiffness was used for the liquefied soil layers (i.e., from depths of 1 m to 4 m and from depths of 7.5 m to 8.5 m).   In addition, recommendations for passive pressure of liquefied sand using a p-multiplier of 0.1 to reduce ultimate pressure obtained from Reese et al.'s sand p-y curves (e.g., Liu and Dobry 1995; Wilson et al. 1999) were also used in this study for comparison. All the analyses in this paper were conducted using the LPILE Plus 4.0m computer code (Reese et al. 2000).

Pile Groups

Two special considerations incorporated into the analysis of the pile groups were the effect of pile head restraint at the pile cap and the effect of pile groups. The approach used to analyze pile group behavior in this study was adopted from the method proposed by Mokwa (1999) and Mokwa and Duncan (2003). In this method, it is recommended that the piles in a group can be modeled as an equivalent single pile with a flexural stiffness equal to the number of piles in the group multiplied by the flexural stiffness of a single pile within the group. Figure 4 shows a schematic of the numerical model used for the analysis of 4-pile group subjected lateral spreading. The p-multiplier approach was used to reduce the soil stiffnesses for each pile in the group to account for group effects.   The reduced soil spring stiffnesses were then summed to develop the combined soil springs for the pile group. For the soil passive pressure acting on the pile cap, it was modeled using the sand p-y curves (Reese et al. 1974), considering the width of pile cap as a pile diameter. In addition, the pile head boundary condition of the group-equivalent pile was determined by estimating the rotational restraint provided by the piles and pile cap and was represented by a rotational spring as suggested by Mokwa (1999). Finally, the group-equivalent pile, incorporating both the effect of pile head restraint and the effect of pile group behavior was analyzed by imposing the free-field soil movement profile at the boundary end of each soil spring.   The input free-field soil movement profiles were the same as that used in the single pile.

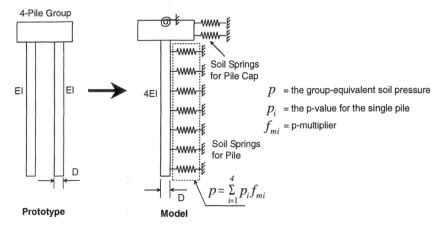

Figure 4: p-y analysis model for pile group

ANALYSIS RESULTS

Single Pile

Figure 5 presents a comparison between computed and measured pile responses of the single pile for the first and second experiments. When using zero spring stiffness for liquefied soil, the predicted pile head displacements (Figure 5a), pile head rotations (Figure 5b) and moment profiles (Figure 5c) were in very good agreement with the pile responses measured from both tests. Figure 5 also shows that using the p-multiplier approach for the liquefied soil layers overestimated the pile displacement, rotation, and moment. Because the behavior of liquefied soil at the test site could be well represented by using zero spring stiffness, only this method was used to analyze the behavior of pile groups in the subsequent sections of this paper.

Figure 5a shows that for the first 8 m, the movement of the free-field soil mass was greater than the movement of the pile, which implies that the soil provided the driving force to the pile, as is also shown by the positive soil reaction in Figure 5d, except for the liquefied layer where zero reaction was assumed. Negative soil resistance indicates that the soil mass moved less than the pile, and therefore the soil provided the resistance force to the pile, as mostly occurred in very dense gravel layer (Figure 5d).

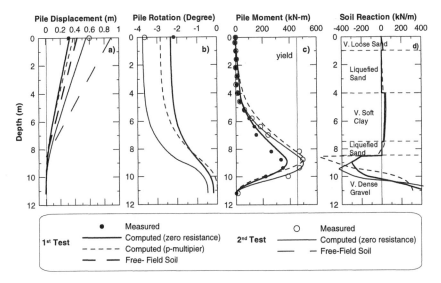

Figure 5: Comparison between measured and computed pile responses for the single pile for 1st and 2nd tests

The lateral responses of the pile are dependent on the magnitude of free-field soil displacement; increasing the free-field soil displacement results in higher maximum moment and pile head displacement. However, if the free-field soil movement is large enough to cause the lateral soil pressure to be fully mobilized, the response of the pile will be independent of the free-field soil movement.   Analyses were conducted to determine the magnitude of surface displacement required to cause the soil to reach its ultimate, or limiting, pressure. The pile stiffness was assumed to remain linear-elastic throughout the analyses. The analyses were carried out by gradually increasing the ground surface displacement while the soil displacement profile was assumed to be linear as shown in Figure 3. Figure 6a shows that once the ground surface displacement is larger than 1.4 m, the laterally spreading soil reaches its ultimate pressure, resulting in no change in the pile head displacement and pile moment. The analysis results indicate that the maximum moment (537 kN-m) in the pile, assuming linear pile behavior, is greater than the yield moment (460 kN-m); therefore, the pile will yield before the soils reach the ultimate pressure, which was indeed observed from pile data in the second test. Additional analyses were then conducted using actual nonlinear pile properties. Figure 6b shows the analysis results of a nonlinear pile which indicates that the pile yields when the soil surface displacement reaches 0.65 m. Because of the pile yielding, increasing the ground displacement increases the pile head displacement. The measured pile displacements and maximum moments from both tests are also plotted in Figure 6b, indicating good agreement between the p-y analysis method and the test results for the single pile.

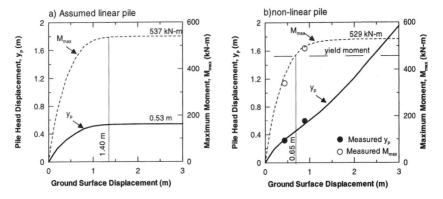

Figure 6:   Pile head displacement and maximum moment vs. ground surface displacement for UCSD single pile; a) assumed linear pile behavior; b) non-linear pile behavior

Pile Groups

Although observations from the Tokachi experiments (Ashford and Juirnarongrit 2004) indicated that small rotations might have occurred at the connection between the pile heads and the pile cap due to the cracking of the concrete around the embedded piles and the elongation of the anchor bars, this effect was assumed to be small, especially for the first test (i.e., the maximum negative moment at the pile head was significantly lower than the ultimate moment capacity at the connection). As a result, the rotations of the cap and the pile heads were assumed to be identical. However, for the second test, this assumption might not be valid and the effect of the difference in the rotations between the cap and the pile heads has to be taken into consideration. Due to the limitation of incorporating this effect into the numerical model used in this study, the analyses were only conducted to predict the results from the first experiment and are presented below.

Figure 7 and Figure 8 present the results of calculated and measured pile responses of the 4-pile group and the 9-pile group, respectively. The same soil properties and free-field soil displacement profile used in the case of single pile were also used for analyzing the behavior of pile groups. Three types of boundary conditions at the pile head were considered for the purpose of comparison; these include the free head condition, fixed head condition, and rotationally restrained pile head boundary condition. Neither the free head nor fixed head conditions provided a reasonable estimate of the measured pile behavior. The free-head case overestimated the maximum positive moment at depth, and gave zero moment at the pile head, while the fixed-head case under-predicted the maximum positive moment but overestimated the maximum negative moment. The deflections at the pile head obtained from the fixed-head case were smaller than that measured by approximately 50% for both the 4- and 9-pile groups. The free-head case over-predicted the pile head deflection by 53% and 60% for the 4-pile group and the 9-pile group, respectively.

The analysis results obtained using the rotationally restrained pile head boundary condition considerably improved the agreement between measured and computed responses for both 4-pile group and 9-pile group, as shown in Figure 7 and Figure 8, respectively. The computed pile moments were within a reasonable range of the measured moment from the test. The errors between computed and measured pile group displacements were 3% for the 4-pile group and 13% for the 9-pile group. Pile head rotation was somewhat overestimated on the 9-pile group as shown in Figure 7b, likely due to the difference in the amount of rotation between the pile heads and the pile cap.

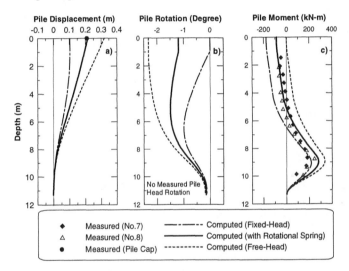

Figure 7: Comparison between measured and computed pile responses for 4-pile group for 1st test

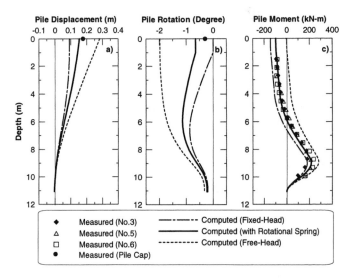

Figure 8:   Comparison between measured and computed pile responses for 9-pile
group for 1st test

CONCLUSIONS

In this study, the push-over analysis using p-y approach was used to predict the
behavior of single pile and pile groups subjected to lateral spreading using a single set
of baseline soil properties. The analysis results were compared to the results from the
full-scale lateral spreading tests in the Port of Tokachi, Japan. Responses of the
single piles subjected to lateral spreading were determined by imposing the known
free-field soil movement profile measured during the tests to the boundary condition
of Winkler springs. Standard p-y springs available in the literature were used to
model the stiffnesess of non-liquefied soil layers, while no soil resistance was used
for the liquefied soil layer. For the case of pile groups, the piles in the groups were
modeled as an equivalent single pile with a flexural stiffness equal to the number of
piles in the group multiplied by the flexural stiffness of a single pile within the group.
The rotational spring and a decrease of soil spring stiffnesses using p-multiplier
approach were incorporated into the pile group model to account for the rotational
pile group stiffness and the pile group effect, respectively. Then, the analyses for the
case of pile group can be conducted in similar way as the single pile case. Computed
pile responses for each pile foundation were compared to the measured responses
obtained from the tests. Reasonably good agreement for all types of pile foundations
considered in this study was obtained between the computed and measured responses.
The results provide justification for the use of push-over analysis using p-y approach
in piles subjected to lateral spreading problems.

ACKNOWLEDGMENTS

This research was sponsored by the PEER Lifelines Program with support from Caltrans, Pacific Gas & Electric and the California Energy Commission under Contract No. 65A0058, as well as by the Pacific Earthquake Engineering Research Center under NSF Contract No. EEC-9701568.

REFERENCES

Ashford, S. A., and Juirnarongrit, T. (2004). "Performance of lifelines subjected to lateral spreading," *Report No. SSRP-04/18*, Department of Structural Engineering, University of California, San Diego.

Ashford, S. A., and Rollins, K. M. (2002). *TILT: Treasure Island Liquefaction Test Final Report*, Report No. SSRP-2001/17, Department of Structural Engineering, University of California, San Diego.

Boulanger, R. W., Kutter, B. L., Brandenberg, S. J., Singh, P., and Chang, P. (2003). *Pile Foundations in Liquefied and Laterally Spreading Ground during Earthquakes: Centrifuge Experiments and Analyses*, Report No. UCD/CGM-03/01, Department of Civil and Environmental Engineering, University of California at Davis.

Dobry R., Abdoun, T., O'Rourke, T., and Goh, S. H. (2003). "Single piles in lateral spreads: Field bending moment evaluation." *Journal of Geotechnical and Geoenvironmental Engineering*, ASCE, 129 (10), 879-889.

Japanese Road Association (JRA). (2002). *Specifications for Highway Bridges*, Japan Road Association, Preliminary English Version, prepared by Public Works Research Institute (PWRI) and Civil Engineering Research Laboratory (CRL), Japan, November.

Liu, L. and Dobry, R. (1995). "Effect of liquefaction on lateral response of piles by centrifuge model tests." *NCEER Bulletin*, Vol. 9, No. 1, 7-11.

Matlock, H. (1970). "Correlations for design of laterally loaded piles in soft clay." *Proc., 2nd Annual Offshore Technology Conf.*, Paper No. OTC 1204, Houston, Texas, 577-594.

Meyersohn, W. D. (1994). "Pile response to liquefaction induced lateral spread." Ph.D. thesis, Dept. of Civil and Environmental Engineering, Cornell University, Ithaca, N.Y.

Mokwa, R. L. (1999). "Investigation of the resistance of pile caps to lateral spreading." Ph.D. thesis, Dept. of Civil Engineering, Virginia Polytechnic Institute and State University, Blacksburg, Virginia.

Mokwa, R. L., and Duncan, J. M. (2003). "Rotational restraint of pile caps during lateral loading." *Journal of Geotechnical and Geoenvironmental Engineering*, ASCE, 129(9), 829-837.

Reese, L. C., Cox, W. R., and Koop, F. D. (1974). "Analysis of laterally loaded piles in sand." *Proc. 6th Offshore Technology Conference*, Paper 2080, Houston, Texas, 473-483.

Reese, L. C., Wang, S. T., Isenhower, W. M., and Arrellaga, J. A. (2000). *Computer Program LPILE Plus Version 4.0 Technical Manual*, Ensoft, Inc., Austin, Texas.

Seed, H. B., and Idriss, I. M. (1971). "A simplified procedure for evaluating soil liquefaction potential." *JSMFD*, ASCE, 97(9), 1249-1274.

Tokimatsu, K., Suzuki, H., and Suzuki, Y. (2001). "Back-calculated p-y relation of liquefied soils from large shaking table tests." *Fourth International Conference on Recent Advances in Geotechnical Earthquake Engineering and Soil Dynamics*, S. Prakash, ed., University of Missouri-Rolla, paper 6.24.

Tokimatsu, K., and Asaka, Y. (1998). "Effects of liquefaction-induced ground displacements on pile performance in the 1995 Hyogoken-Nambu earthquake." *Soils and Foundations*, Special Issue on Geotechnical Aspects of the January 17, 1995 Hyogoken-Nambu Earthquake, 163-177.

Wilson, D. W., Boulanger, R. W., and Kutter, B. L. (1999). "Lateral resistance of piles in liquefying soil." *OTRC'99 Conf. On Analysis, Design, Construction & Testing of Deep Foundations*, J. M. Rosset, ed., Geotechnical Special Publication No. 88, ASCE, 165-179.

Wilson, D. W., Boulanger, R. W., and Kutter, M. L. (2000). "Observed seismic lateral resistance of liquefying sand." *Journal of Geotechnical and Geoenvironmental Engineering*, ASCE, 126(10), 898-906.

DAMAGE OF PILES CAUSED BY LATERAL
SPREADING – BACK STUDY OF THREE CASES

San-Shyan Lin, Member, ASCE[1], Yu-Ju Tseng[2], Chen-Chia Chiang[2] and C.L. Hung[3]

ABSTRACT

Three case histories of failure of piles caused by ground lateral spreading are back studied using the beam on Winkler foundation method including nonlinear soil springs and a nonlinear moment-curvature relation for the pile. Soil loads on the pile are modeled based on the modified Bouc-Wen model. The lateral spreading effect is modeled by laterally displacing the soil spring support by an amount equal to the free field permanent deformation. Design procedures suggested by Tokimatsu et al. (1998) and by JRA (1996) were also used for case histories evaluation and compared to available observation results. Buckling analysis was performed to determine critical buckling load. Whether the cause of failure for the evaluated cases was due to either bending or buckling is also discussed in the paper.

---

[1] Professor, Dept. of Harbor and River Eng., National Taiwan Ocean University, Keelung, Taiwan 20224.
[2] Graduate Student, Dept. of Harbor and River Eng., National Taiwan Ocean University, Keelung, Taiwan 20224.
[3] Assistant Researcher, Taiwan Construction Research Institute, Taipei County, Taiwan.

INTRODUCTION

Lateral spreads were observed after many well-known earthquakes such as the 1964 Niigata earthquake (Hamada 1992), the 1995 Kobe earthquake (Tokimatsu 2003) and the 1999 Chi-Chi earthquake (Hwang et al. 2003) had left extensive damage to many pile foundations of bridges and buildings. Excavations performed at some buildings in Niigata City 20 years after the earthquake or field survey performed at some building in Kobe City showed bending failure of piles occurred at the boundaries of liquefied and non-liquefied soil. It was suggested the pile might have been damaged due to liquefaction induced permanent ground displacement. Subsequently, the effects of ground movements on piles were suggested to properly taken into account in pile foundation design. The Japanese Highway Code of Practice (JRA 1996) has codified this concept after the 1995 Kobe earthquake.

Bhattacharya et al. (2004) proposed an alternative mechanism of pile failure in liquefiable deposits during earthquakes. It was considered that the pile becomes unstable under axial load from loss of support from the surrounding liquefied soil, provided the slenderness ratio of the pile in the unsupported zone exceeds a critical value. The instability causes the pile to buckle and causes a plastic hinge in the pile. In terms of soil pile interaction, the method assumes that, during instability, the pile pushes the soil. Hence, the lateral load effects are considered to be secondary to the basic requirement that piles in liquefiable soils must be checked against Euler's buckling.

However, in 1994, it was already proposed by Meyersohn (1994) that piles subject to lateral spreads resulting from soil liquefaction might cause two distinct failure modes. The first one is lateral pile deflections induced by ground lateral spreads that may result in the pile reaching its bending capacity and hence develop a plastic hinge. Another failure mode is the combined action of lack of sufficient lateral support due to the reduced stiffness of the liquefied soil and the lateral deflection imposed on the pile, may result in pile buckling. Whether bending or buckling mode of a pile may develop depends primarily on the stiffness of the liquefied soil, length of pile exposed to liquefied soil, axial load imposed to pile, and bending stiffness of the pile.

Although failure of piles caused by ground lateral spreads may be due to combined and simultaneous actions of permanent ground displacement effect and axial loading effect, separate analyses for either bending or buckling failure pattern studies are often adopted by the geotechnical engineers. The purpose of this paper is to study the possible failure modes of three available pile failure cases. Whether these piles were failed by either bending or buckling mode was re-evaluated. In addition, design procedures suggested by Tokimatsu et al. (1998) and by JRA (1996) were also used for case histories evaluation and compared to available observation results. The nonlinear pile soil interaction is modeled as pile with Winkler foundation interaction, via continuous distribution of nonlinear springs. The Bouc-Wen model (Wen 1985) is used to simulate the soil and pile material behaviors. These models as well as beam-column finite elements were implemented into the available computer program

SPASM (Matlock et al. 1978). One more benefit of adopting coding structure of the program SPASM is the availability of buckling analysis already coded in the program.

MATERIAL MODELING

The Bouc-Wen model was successfully used by the first author and other colleagues for nonlinear static and dynamic analysis of concrete piles (Lin et al. 2001 and 2002). The material models are briefly reviewed in the following section. Detail description of the Bouc-Wen model can be referred to Wen (1985) or Lin et al. (2001 and 2002).

Soil Modeling

Based on the Bouc-Wen model, the force resulting from the nonlinear spring alone can be given as (Wen 1985)

$$F_s(z) = \alpha \cdot K \cdot y + (1 - \alpha) \cdot K \cdot y_0 \cdot \varsigma \tag{1}$$

where α = a parameter controls the post yielding stiffness; K = a reference stiffness; y = the pile deflection at the location of the spring; y_0 = the value of pile deflection that initiates yielding in the spring; and ς = a hysteretic dimensionless quantity that is governed by the following equation (Wen 1985)

$$y_0\dot\varsigma + \gamma \cdot |\dot y| \cdot \dot\varsigma \cdot |\varsigma|^{n-1} + \beta \cdot \dot y \cdot |\varsigma|^n - \dot y = 0 \tag{2}$$

where β, γ and n are parameters that control the shape of the hysteretic loop and are chosen such that the shape of the loop are reasonable for the type of material considered. Eliminating the time variable in Eq. (2) we obtain

$$y_0 \frac{d\varsigma}{dy} = 1 \pm (\beta \pm \gamma)\varsigma^n \tag{3}$$

in which the ($\pm$) sign depends on whether the values of $\dot y$ and ς are positive or negative. The maximum value of ς is given as

$$\varsigma_{max} = (\frac{1}{\beta + \gamma})^{1/n}, \text{ when } d\varsigma / dy = 0 \tag{4}$$

In addition, the maximum value of the spring reaction is reached when y and ς take their maximum value. Hence, for the virgin initial loading, $F_{s,max}$ is reached at the first loading-reversal point ($d\varsigma / dy = 0$)

$$F_{s,\max} = \alpha \cdot K \cdot y_{\max} + (1-\alpha) \cdot K \cdot y_0 (\frac{1}{\beta + \gamma})^{1/n} \qquad (5)$$

Based on the research by Broms (1964), the ultimate value of $F_{s.\max}$ for cohesionless soils is given as

$$F_{s,\max}(z) = \mu \gamma_s d \frac{1 + \sin \phi_s}{1 - \sin \phi_s} z \qquad (6)$$

in which γ_s in the specific weight of the soil; d = the pile diameter; ϕ_s = the internal friction angle of the soil; z = vertical distance from the ground surface; and μ = a parameter which is between 3 and 5 for cohesionless soils (Badoni 1997).

With $\alpha = 0$, $\beta + \gamma = 1$ and the combination of (5) and (6), the spring reactions of the pile for cohesion-less soils were given by Badoni and Makris (1996) as

$$F_s(z) = \mu \cdot \gamma_s \cdot d \cdot \frac{1 + \sin \phi_s}{1 - \sin \phi_s} \cdot z \cdot \zeta \qquad (7)$$

In addition, y_0, or the value of pile deflection at which yielding initiates in the spring can be given as (Badoni and Makris 1996)

$$y_0(z) = \mu \frac{\gamma_s d}{K(z)} \frac{1 + \sin \phi_s}{1 - \sin \phi_s} z \quad \text{(cohesion-less soil)} \qquad (8)$$

Moment-Curvature Relation of the Pile

The Bouc-Wen model is also used to model moment-curvature relationship of the pile and is expressed as (Lin et al. 2001)

$$M = \alpha_M (E_P I_P)\kappa + (1 - \alpha_M)M_y \varsigma_M \qquad (9)$$

where M = the moment; M_y = the yield moment; κ = the curvature; α_M = a parameter controlling the rigidity of the pile after yielding; and ς_M = the hysteretic parameter, which can be expressed as (Lin et al. 2001)

$$\dot{\varsigma}_M = \left\{ A_M I_P - \left[B_M \cdot \varsigma^2 \{Sgn(\dot{\kappa} \cdot \varsigma_M) + 1\} \frac{1}{\kappa_y} \right] \right\} \dot{\kappa} \qquad (10)$$

in which $Sgn(\dot{\kappa} \cdot \varsigma_M) = 1$ if $\dot{\kappa} \cdot \varsigma_M > 0$; $Sgn(\dot{\kappa} \cdot \varsigma_M) = -1$ if $\dot{\kappa} \cdot \varsigma_M < 0$; κ_y = the yielding curvature; and A_M and B_M = the parameters controlling the shape of the hysteretic loop.

For concrete piles, once the moment induced on the pile exceeds a certain magnitude, the moment of inertia of the pile may be reduced due to concrete cracking. A semi-empirical moment versus moment of inertia relationship is used in this paper (Lin et al. 2001) and is given as

$$I_{ef} = I^{I}, \left(M < M_{cr}\right) \tag{11}$$

$$I_{ef} = I^{II} + \left(I^{I} - I^{II}\right)\left(\frac{M_{cr}}{M}\right)^{3}, \left(M_{cr} < M < M_{u}\right) \tag{12}$$

where I_{ef} = the effective moment of inertia; $I^{I} = I_{p}$, the moment of inertia of the non-cracked section; I^{II} = the moment of inertia of the completely cracked section where the reinforcement has reached the yield strength; M_{cr} = the bending moment corresponding to the beginning of cracking; and M_{u} = the bending moment corresponding to I^{II}.

In order to take into account the effects of finite size of plastic regions, the model chosen for this study is based on the global frame member model proposed by Roufaiel and Meyer (1987), in which the model was also successfully used for concrete pile analyses (Badoni 1997).

GOVERNING EQUATION OF THE PILE-SOIL SYSTEM

For force based analysis, such as the method used by JRA (1996), the equilibrium of the pile-soil system can be expressed as

$$E_{P}I_{P}\frac{d^{4}}{dz^{4}}y(z) + \alpha K y(z) + (1-\alpha)K y_{0}\varsigma(z) = 0 \tag{13}$$

However, for displacement based analysis such as the method proposed by Tokimatsu et al. (1998), the equilibrium of the pile-soil system is expressed in the following form as

$$E_{p}I_{p}\frac{d^{4}}{dz^{4}}y(z) + \alpha K \varepsilon(z) + (1-\alpha)K \varepsilon_{0}\varsigma(z) = 0 \tag{14}$$

in which $\varepsilon(z) = y(z) - f_{LS}(z)$, where f_{LS} = the estimated or the measured ground displacement resulted from ground lateral spreading.

CASE STUDIES

Case I - Yachiyo Bridge, Japan (Hamada 1992)

The abutment and piers of the Yachiyo Bridge was damaged during the 1964 Niigata earthquake. The foundations of the abutment and the two piers next to the abutment were sitting on reinforced concrete piles with a diameter of 300mm and a length of about 10m. Fig. 1 shows one of the damaged piles that was extracted and examined after the earthquake. It was reported that the pile was severely destroyed at a depth of about 8m from the top of the pile. In addition, horizontal cracks caused by large bending moment were found. The permanent ground displacement on both banks of the river was found to be 4 to 6m toward the river. Investigations by Hamada (1992) concluded the foundations of the piers were pushed toward the river due to large ground displacement.

Table 1. Material properties used for case study analyses

Back studied cases	Case I (Yachiyo Bridge)	Case II (4-Story Building)	Case III (Showa Bridge)
Pile			
$E_p I_p$ (kN-m^2)	5000	5000	56300
M_{cr} (kN-m)	9	12	1000
M_y (kN-m)	41	75	1000
M_u (kN-m)	50	100	1000
κ_y (1/m)	0.0082	0.015	0.017
Soil			
Upper Non-liquefied layer			
depth (m)	0.0-3.0	0.0-1.5	-
γ_s (kN/m^3)	14.4	16.5	-
ϕ_s (deg.)	30	32	-
Middle(or upper) liquefied layer			
depth (m)	3.0-11.0	1.5-14.0	0.0-19.0
γ_s (kN/m^3)	12.5	12.5	16.5
ϕ_s (deg.)	25	25	33
Bottom non-liquefied layer			
depth (m)	11.0-14.0	14.0-23.0	19.0-25.0
γ_s (kN/m^3)	18.5	18.5	18.5
ϕ_s (deg.)	40	40	34

The moment versus curvature properties of the damaged pile is given in Table 1. Results for the studies of the pile displacements based on the JRA (1996) procedures and the Tokimatsu et al method (1998) are shown in Fig. 2, in which results estimated from both methods are slightly higher than the observed pile displacements.

A scaling factor for the horizontal subgrade reaction of 0.06 was used for liquefied soils throughout the study of this case. However, all the predicted pile bending moments have reached yield moments of the pile at boundaries of the liquefied and non-liquefied soil layers. The predicted critical buckling load of the pile is about 250kN, as given in Fig. 3.

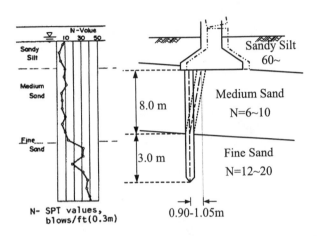

Figure 1 : Damage of a foundation pile at Yachiyo bridge (after Hamada 1992)

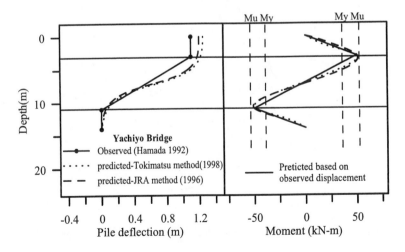

Figure 2 : Analytical results for piles at the Yachiyo bridge

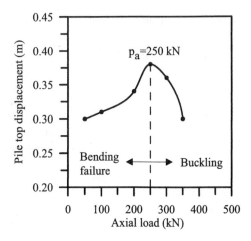

Figure 3 ： Maximum soil displacement versus axial load for pile at Yachiyo bridge

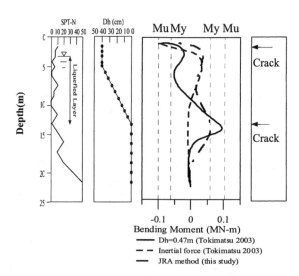

Figure 4 ： Computed bending moment of the Case II (after Tokimatsu 2003)

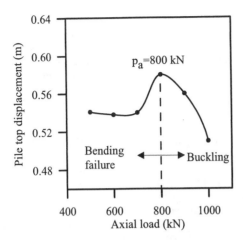

Figure 5 : Maximum soil displacement versus axial load for pile at Case II

Case II – A 4-Story Building in Mikagehoma, Japan (Tokimatsu 2003)

A case history of 35cm diameter and 23m long pre-stressed high strength concrete piles supporting a four-story building in Mikagehoma was reviewed by Tokimatsu (2003) after the 1995 Kobe earthquake. Field survey showed that the piles cracked near the pile head and near the bottom of the fill causing tilting of the building, as shown in Fig. 4, in which the distribution of computed bending moments with depth is also included. A scaling factor for the horizontal subgrade reaction of 0.06 was used for liquefied soils. The predicted maximum moment locations match very well the observed concrete crack locations of the pile. The values of the moment obtained from this study are in general higher than that of the predicted values by Tokimatsu (2003), especially at the upper portion of the pile that is closer to the yield moment of the pile. Although the actual axial loading carried by the piles is unknown, the critical buckling load for this particular case is estimated to be 800kN, as shown in Fig.5.

Case III- Showa Bridge, Japan (Hamada 1992; Bhattacharya et al. 2004 and 2002)

The 12-span, 307m long Showa Bridge failed during the 1964 Niigata earthquake. Five simply supported steel girders each with 28m span, fell into the river. The piers were constructed by driving steel pipe piles. Investigation was conducted after the earthquake and recovery of the damaged pile along with the soil condition is given in Fig.6, in which the deformed pile was extracted after the earthquake. The pile, with a diameter of 60.9cm and a thickness of 0.18cm, was bent toward the right bank of the river at a point 7 to 8m below the riverbed. The liquefied layer, on the left bank of the river, was estimated to be about 10m thick, slid toward the center of the river by about 5m. Studies by Bhattacharya et al. (2004) suggested that failures of the piles of this particular case might resulted from pile buckling, and the estimated buckling load is 1095kN.

Applying the method proposed by Tokimatsu et al. (1998) and the design procedures given by the Japanese Road Association (1996), the estimated pile displacements are shown in Fig. 7. A scaling factor for the horizontal subgrade reaction of 0.13 was used for liquefied soils. The predicted pile displacements obtained from both methods appear to be smaller than the observed lateral displacement of the pile. The bending moments of the pile caused by ground displacements estimated from these two methods as well as from the observed pile displacement are also given in Fig. 7, in which all the values of the estimated moments are much smaller than the ultimate moment of the pile. Study for critical buckling load is shown in Fig. 8. It gives the critical buckling load of 1000kN, which is slightly less than the estimated values given by Bhattacharya et al. (2004).

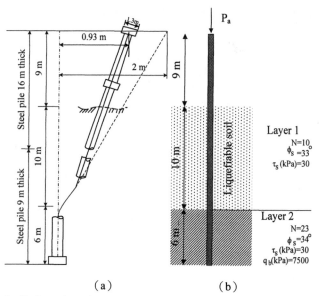

(a)   (b)

Figure 6 : Deformation of a steel pipe pile of one of the piers of Showa bridge and soil information used for analysis (after Bhattacharya et al 2002)

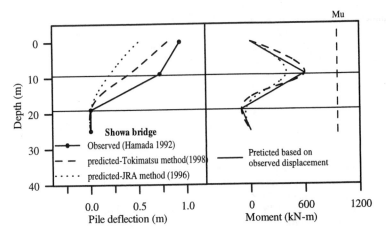

Figure 7 : Analytical results for piles at Showa bridge case

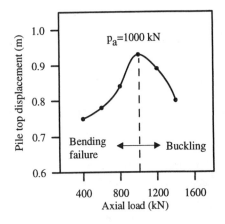

Figure 8 : Maximum soil displacement versus axial load for pile at Showa bridge

SUMMARY AND CONCLUSIONS

Back studies on evaluation of possible failure mode for three failure cases of piles caused by lateral spreads have been presented. The Bouc-Wen model, which had previously been successfully used to represent static and cyclic soil properties for pile analysis (Lin et al. 2001; and Badoni and Makris 1996), has been extended here to model the effects of concrete cracking on the pile performance caused by ground lateral spreads. Critical buckling load of the piles was also calculated for the evaluated cases. In addition, design procedures suggested by Tokimatsu et al. (1998) and by JRA (1996) were also used for case histories evaluation and compared to available observation results. Whether the evaluated pile cases failed because of bending or buckling is also discussed. For the three case histories evaluated herein, the following conclusions may be drawn:

1. Only for Case III, the Showa Bridge case, was the failure of the pile due to buckling. Whereas, pile damage of the other two cases was due to bending failure.
2. The predicted ground lateral spread effects on pile displacements using the procedures proposed by Tokimatsu et al. (1998) are larger than that using JRA (1996) method. Subsequently, the former method predicts higher pile bending moments than the later one.
3. The buckling load predicted from this study is slightly lower than the result obtained by Bhattacharya et al. (2004).

ACKNOWLEDGEMENT

This study was support by the National Science Council and by the Ministry of Transportation and Communications, Taiwan, ROC, under contract number NSC-92-2211-E-019-007 and MOTC-STAO-93-05, respectively.

REFERENCES

Badoni , D., and Makris, N. (1996). "Nonlinear response of single piles under lateral inertial and seismic loads." *Soil Dynamics and Earthquake Engineering.*, 15, 29-43.

Badoni, D. (1997). "Nonlinear response of pile foundation-superstructure systems." Doctoral Dissertation, Department of Civil Engineering, *University of Notre Dame, Indiana, USA.*

Bhattacharya, S., Madabhushi, S.P.G., and Bolton, M.D. (2002). "An alternative mechanism of pile failure in liquefiable deposits during earthquakes." *Technical Report CUED/D-SOILS/TR324*, University of Cambridge, UK. http://www-civ.eng.cam.ac.uk/geotech_new/publications/TR/TR324.pdf.

Bhattacharya, S., Madabhushi, S.P.G., and Bolton, M.D. (2004). "An alternative mechanism of pile failure in liquefiable deposits during earthquakes." *Geotechnique.*, 54(3), 203-213.

Broms, B.B. (1964) "Lateral resistance of piles in cohesionless soils" *Journal of Soil Mechanics and Foundations Division,* ASCE, 90(SM3), 123-56.

Hamada, M. (1992). "Large ground deformations and their effects on lifelines: 1964 Niigata earthquake" in *Case Studies of Liquefaction and Lifeline Performance During Past Earthquakes, Japanese Case Studies,* Technical Report, NCEER-92-0001, NCEER, Buffalo, NY, USA., Vol. (1), 3.1-3.123.

Hwang, J.H., Yang, C.W., and Chen, C.H. (2003). "Investigations on soil liquefaction during the Chi-Chi earthquake." *Soils and Foundations.*, 43(6), 107-123.

Japanese Road Association (1996). "Specifications for highway bridges", *Part V. Seismic Design.*

Lin, S.S., Liao, J.C., Liang, T.T., and Juang, C.H. (2002). "Use of Bouc-Wen model for seismic analysis of concrete piles." *in Deep Foundations 2002.*, GSP 116, ASCE, 372-384.

Lin, S.S., Liao, J.C., Yang, T.S., and Juang, C.H. (2001). "Nonlinear response of single concrete piles." *Geotechnical Engineering Journal.*, 32(3), 165-175.

Matlock, H., Foo, S.H.C., Tsai, C.F., and Lam, I. (1978). "SPASM 8: A dynamic beam-column program for seismic pile analysis with support motions." *a report to Chevron Oil Field Research Co.*, Fugro, Inc., Long Beach, CA, USA.

Meyersohn, W.D. (1994). "Pile response to liquefaction-induced lateral spread." *Doctor's dissertation.*, Cornell University, NY, USA.

Roufaiel, M.S.L., and Meyer, C. (1987). "Analytical modeling of hysteretic behavior of R/C frames." *Journal of Structural Engineering.*, ASCE, 113(3), 429-444.

Tokimatsu, K. (2003). "Behavior and design of pile foundations subjected to earthquakes." *Keynote speech, in Proceedings 12th Asia Regional Conference on Soil Mechanics and Geotechnical Engineering*, Singapore, Vol. (2), 1065-1096.

Tokimatsu, K., Oh-oka, H., Satake, K., Shamoto, Y., and Asaka, Y. (1998). "Effects of lateral ground movements on failure patterns of piles in the 1995 Hyogoken-Nambu earthquake." in *Geotechnical Earthquake Engineering and Soil Dynamics.*, Vol. (3), 1175-1186.

Wen, Y.K. (1985). "Response and damage of hysteretic systems under random excitation." *Proceedings of 4th International Conference on Structural Safety and Reliability*, Vol. (1), 291-300.

STIFFNESS OF PILES IN LIQUEFIABLE SOILS

Pedro Arduino, Member, ASCE[1], Steven L. Kramer, Member, ASCE[2], Ping Li[3], and John C. Horne, Associate Member, ASCE[4]

ABSTRACT

This paper describes the results of an investigation of the dynamic stiffness of pile foundations in liquefiable soils. Such soils frequently exist near bodies of water where bridges are required. Because potentially liquefiable soils are generally weak and compressible even under static conditions, bridges founded on them are usually supported on pile foundations. During earthquake shaking, the excess porewater pressure that builds up in liquefiable soils influences both the seismic response of the soil deposit itself, and also the local interaction with any foundations that extend through the liquefiable soil. These phenomena can strongly influence the stiffness of pile foundations in liquefiable soils during earthquakes, and need to be accounted for if accurate evaluations of pile stiffness are to be obtained. In this work, tools and procedures were developed for evaluation of the stiffness of pile foundations in liquefiable soils during earthquakes.

INTRODUCTION

Because they frequently cross bodies of water such as rivers, streams, and lakes, highway bridge foundations are often supported on fluvial and alluvial soil deposits that contain loose, saturated sands and silty sands. These deposits are generally weak

---

[1]J. Ray Bowen Professor for Innovation in Engineering Education, University of Washington, Seattle, WA.
[2]John R. Kiely Professor of Civil and Environmental Engineering, University of Washington, Seattle, WA.
[3]Graduate Student, Department of Statistics, Stanford University, Stanford, CA.
[4]Supervising Engineer, Parsons Brinckerhoff Quade & Douglas, Portland, OR.

and/or soft enough that deep foundations are required to support the bridge without excessive settlement or without damage due to phenomena such as erosion and scour. In many cases, bridges are supported on groups of driven piles; in some cases, drilled shaft foundations may be used to support bridges. These foundations extend through relatively shallow deposits of loose soil to derive their support from deeper and/or denser soils.

In seismically active environments, loose, saturated soil deposits are susceptible to liquefaction. As porewater pressures increase in a liquefiable soil, the effective stress decreases and with it the strength and stiffness of the soil are reduced. The level of the reduction may be modest, or it may be considerable. As the stiffness of liquefiable soil changes, the resistance it can provide to the movement of foundation elements supported in it also changes. As a result, the stiffness of foundations supported in or extending through liquefiable soils will change during earthquake shaking. Because the structural response of a bridge depends on the stiffness of the bridge foundation, accurate estimation of structural response requires accurate estimation of foundation stiffness. Therefore, the reliable design of new bridges, or seismic evaluation of existing bridges, in liquefiable soil requires accurate characterization and modeling of foundation stiffnesses.

In liquefiable soil conditions, characterization of foundation stiffness requires prediction of free-field soil response (response of the soil in the absence of a foundation) and prediction of soil-pile interaction behavior. Because the liquefaction process is complicated and time-dependent, the interaction between soil and deep foundations is difficult to account for. As a result, characterization of pile foundation stiffness in liquefiable soil has been a difficult problem for geotechnical engineers.

Background

To simplify the process of evaluating the stiffnesses of foundations for foundation types and soil conditions commonly encountered in Washington state, the Washington State Department of Transportation (WSDOT) contracted with a consultant to develop a series of procedures and charts for estimating pile stiffness. These procedures and charts were presented in a *Design Manual for Foundation Stiffness Under Seismic Loadings* (Geospectra, 1997), which will be referred to as the "*Manual*" in the remainder of this paper.

The *Manual* included procedures and charts for estimating the stiffness of deep foundations in liquefiable soil conditions. However, it noted that these procedures were based on a series of assumptions that greatly simplified a complicated process and, therefore, were very approximate. Improved understanding of the process of soil liquefaction and of soil-pile interaction, and the development of improved procedures for modeling these phenomena, offer the opportunity to improve the accuracy and reliability of procedures for evaluating the stiffness of deep foundations in liquefiable soils. This paper briefly describes a series of analyses performed to provide improved estimates of the stiffness of pile foundations in liquefiable soils.

Case Histories of Liquefaction-Induced Pile Damage

The damaging effects of soil liquefaction on pile foundations and the structures they support have been observed in past earthquakes and reproduced in laboratory model tests. A brief review of some of these observations helps to illustrate the phenomena involved and to identify the important aspects of soil and foundation behavior that must be considered in a foundation stiffness analysis.

Pile foundations can be damaged not only by excessive loads transmitted to the pile head from the structure, but also by non-uniform lateral soil movements. Such movements, induce bending moments and shear forces in piles. The damaging effects of lateral soil movements on pile foundations are well documented from past earthquakes. For example, in most cases involving lateral spreading the majority of the observed pile damage can be attributed to the horizontal loads applied to the piles by the laterally spreading soil. Such damage has been observed in several past earthquakes. Lateral spreading displacement of about 1 m in the 1964 Niigata earthquake caused severe damage to piles supporting the NHK building (Figure 1).

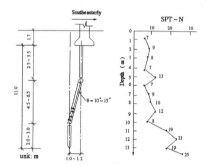 

Figure 1. Schematic illustration (left) and photograph (right) of piles damaged by lateral spreading in the 1964 Niigata earthquake (Hamada, 1992).

Softening of pile foundations in liquefiable soils, in combination with forces caused by soil movements, has also caused substantial damage to bridges. In the 1995 Hyogo-ken Nanbu earthquake in Kobe, Japan, a span of the Nishinomiya bridge (Figure 2) fell to the ground; the fact that the distance between the supports was shorter than the length of the fallen span indicates that the foundation stiffness was low enough to allow large dynamic deflections and/or rotations of the pile foundations supporting the bridge.

Figure 2. Nishinomiya Bridge following 1995 Hyogo-ken Nanbu earthquake.

Simplified Procedure

The *Manual* presented simple procedures for estimating foundation stiffnesses of typical bridge foundations for three different ground shaking levels (PGA values of 0.2 g, 0.3 g and 0.4 g). The *Manual* presented normalized stiffness curves for all six degrees of freedom (three translational and three rotational) for a total of 19 standard foundation scenarios generated from seven standard soil profiles and six typical pile foundation configurations. The scenarios were considered to represent many of the typical bridge foundations in the State of Washington.

To generate the stiffness charts, one-dimensional, equivalent linear site response analyses were performed for all soil profiles and for three shaking levels (0.2 g, 0.3 g, and 0.4 g) with the computer program SHAKE. SHAKE provided strain-compatible, free-field soil properties (damping, secant stiffness and shear wave velocity), which accounted for the cyclic degradation of soil properties due to nonlinearity effects. The input motion was the 1949 Olympia Earthquake, which was fitted and scaled to match the target spectra.

Two of the seven standard foundation scenarios included liquefiable soil. Soil profile S5 consisted of 3 m of loose, saturated sand overlying 15 m of soft organic silt, which was underlain by dense sand. Soil profile S7 consisted of 3 m of dry medium dense sand overlying 12 m of loose, saturated sand, which was underlain by dense to very dense glacial deposits. The *Manual* presented stiffness curves for one foundation type (P5) in soil profile S5 and two foundation types (P1 and P4) in soil profile S7. Because foundation type P5 had a 3 m-thick footing (pile cap) that extended to the bottom of the liquefiable layer, its stiffness was not influenced by the interaction between piles and liquefiable soils; as a result, soil profile S5 and foundation type P5 are not further considered in this paper. From the other two foundations, foundation type P1 is a 30 m long cast-in-place concrete shaft with external diameters varying from 1.5 m to 2.4 m and a 0.0191 m thick steel casing in the upper 9 meters. The pile head condition is assumed to be free. Foundation type P4 is a pile group composed of nine 30 m long pipe piles filled with concrete, with external diameters of 0.6 m, and

0.0127 m in wall thickness. In this study, only a single pile of the group is analyzed. Both free-head and fixed-head cases are considered.

The procedure for evaluation of the foundation stiffnesses reported in the *Manual* was as follows:

1. Compute the horizontal, rocking, and coupled (horizontal-rocking) stiffnesses of single piles using the computer program COM624 (Reese and Sullivan, 1980). Apply a series of increasing loads/moments to evaluate the displacement/rotation dependence of the stiffness terms. The potentially liquefiable soils were treated as cohesive soils with cohesion values equal to the estimated (but unspecified) residual strengths of the liquefied soil.
2. Compute the vertical stiffnesses of single piles using the computer program GROUP. Apply a series of increasing loads/moments to evaluate the displacement dependence of the vertical stiffness.
3. For foundation type P4, compute the pile group interaction factors using the finite element computer program SASSI (Lysmer et al., 1981).
4. Compute the horizontal and vertical pile group stiffnesses as the product of the single pile stiffness, the number of piles in the group, and the pile group interaction factor.
5. Calculate the rocking and torsional stiffnesses of the pile group using the vertical and horizontal single pile stiffnesses, respectively.
6. Add the stiffnesses of the pile cap to that of the pile or pile group to obtain the total foundation stiffness as a function of foundation deflection.

This approach involved a number of simplifying assumptions that, in light of recent improvements in the understanding of liquefiable soil behavior and pile-soil interaction, could lead to inaccurate estimates of the stiffnesses of pile foundations in liquefiable soil. In particular, the approach does not account for the following:

1. The dynamic nature of the problem – the single pile stiffnesses were based on static analyses with constant loads and constant soil properties.
2. The time-dependent buildup of porewater pressure in the soil – this influences the free-field motion and also the *p-y* behavior of the soil, which will change over the duration of an earthquake.
3. The occurrence of cyclic mobility – dilation-induced spikes in stiffness can lead to pulses of high acceleration and, potentially, to increased bending moments and shear forces after initial liquefaction has occurred.
4. Differences in "liquefiable" soils – the *Manual* implicitly assumed that all liquefiable soils are alike and treated them in a highly simplified manner with unspecified properties.
5. High impedance contrasts – high impedance contrasts can lead to increased dynamic response and increased kinematic loading on pile foundations.

6. Sloping ground conditions – the *Manual* assumed level-ground conditions and therefore did not account for lateral soil movement (lateral spreading) that can influence foundation stiffness and also induce significant kinematic loading in piles.

Because the problem of pile stiffness in liquefiable soils is considerably more complicated than that for non-liquefiable soils, the simplifying assumptions used in the *Manual* must be carefully considered. A careful review of these assumptions revealed that recent improvements in liquefaction and soil-pile interaction modeling can eliminate the need for many of them and lead to more accurate estimation of the stiffnesses of a pile.

Among these improvements, new constitutive models can better represent the behavior of liquefiable soils, including such important phenomena as pore pressure generation, stiffness degradation, and phase transformation behavior. Improved soil-pile interaction analyses can also be performed, including nonlinear, inelastic soil resistance (*p-y* behavior), pore pressure-induced softening, and radiation damping. Coupling the improved constitutive models with nonlinear site response analyses and improved soil-pile interaction analyses offers the opportunity for more accurate prediction of the response of pile foundations in liquefiable soil, including more accurate prediction of the stiffness of pile foundations in such soils.

PILE MODELING

Modeling the stiffness of piles in liquefiable soils requires modeling the free-field seismic response of liquefiable soils and modeling the interaction between pile foundations and the liquefiable soil. Free-field soil response can be modeled using one-dimensional site response analyses; for this investigation, site response was modeled using the nonlinear, one-dimensional computer program WAVE (Horne, 1996) with the *UWsand* constitutive model (Kramer et al., 2002). The *UWsand* model allows WAVE to model the effects of pore pressure generation, including phase transformation effects, for soils of different SPT resistance. Soil-pile interaction was modeled using the computer program, DYNOPILE (Horne, 1996; Horne and Kramer, 1998), which models a single pile as a Winkler beam; DYNOPILE models nonlinear, inelastic pile-soil interaction using *p-y* elements and radiation damping using Nogami-Konagai far-field elements (Nogami et al., 1992). In this investigation, the pore pressures generated in the WAVE analyses were used to modify the *p-y* curves and far-field shear wave velocities at each time step in the DYNOPILE analyses. For this investigation, DYNOPILE was also modified to include a nonlinear single-degree-of-freedom (SDOF) attached to the head of the pile.

Effect of Pore Pressure on Pile Bending Moment

The pile bending moment due to lateral seismic loading is one of a foundation designer's major concerns. If a liquefiable layer exists within the depth of the pile, the large free-field differential displacements between the liquefiable and non-liquefiable layers can cause substantial pile bending moments, which may exceed the pile moment capacity.

Figure 3 shows curves of maximum bending moment for different soil and boundary conditions. Pile bending moments for a soil with $(N_1)_{60}$ = 10 layer are very large. The largest values occur near the layer boundaries, i.e., near the top and bottom of the liquefiable layer. This is not surprising since the free-field soil curvatures are largest near the layer boundaries. It is important to note that, with a "fixed-head" boundary condition, the largest pile bending moment may occur near the pile top.

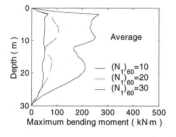 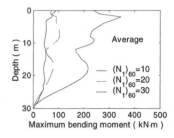

Figure 3. Average envelopes of maximum pile bending moments for P4 foundation at three different soil densities: (a) free head, and b) fixed head conditions.

The instant of time at which the maximum pile bending moment occurs is also of interest. Figure 4 shows three bending moment diagrams picked at different times for the soil density $(N_1)_{60}$=10 and the free-head boundary condition. The maximum bending moment occurs at t = 38.3 sec, which is neither the end of shaking nor the time at which the largest acceleration occurs. The maximum acceleration occurs at t = 21.3 sec. At this time, the excess pore pressures have not built up to a high level, therefore, the soil and pile stiffness have not degraded much. This can be clearly seen in the pore pressure time history shown in Figure 4. After some time, the soil stiffness degrades to a very low value inducing high bending moments along the pile. At the end of shaking, the soil-pile system adjusts the stress distribution and gradually decreases the pile bending moment.

Effects of Coupled vs Uncoupled Structural Response

Soil-pile-structure interaction analyses can be performed in two ways. In an uncoupled analysis, a pile-soil interaction analysis is performed to compute the motion of the top of the pile under earthquake loading. That motion is then applied as a fixed-base motion to a separate structural model. As a result, the response of the structure is not influenced directly by the response of the pile (or vice versa). In a coupled analysis, the response of the structure and pile are computed simultaneously so that the interaction between the two is fully accounted for.

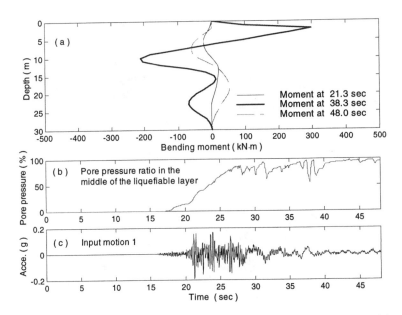

Figure 4. Pile bending moments for foundation type P4 at selected instants of time for $(N_1)_{60} = 10$ liquefiable layer.

Figure 5 shows the uncoupled and coupled response of linear SDOF systems with natural periods of 0.3, 0.5, 1.0, and 1.5 sec to identical motions applied at the base of the soil deposit. These analyses showed substantially higher total displacements in the coupled analyses, particularly for the longer-period structures. The difference between uncoupled and coupled response for similar analyses in which the pile head was assumed fixed against rotation was much smaller. These results indicate that rocking effects can be significant for piles with free-head conditions.

Effects of Linear vs Nonlinear Structural Response

Earthquake-resistant design of structures typically allows inelastic response to occur in structures with sufficient ductility to accommodate that response without loss of load-carrying capacity. A series of coupled analyses were performed to investigate the effects of inelastic structural response on pile demands. Table 1 shows the characteristics of two sets of three structures that were analyzed; the two sets had elastic periods of 0.5 sec and 1.0 sec, respectively. Within each set, one structure was elastic and the other two were inelastic (bilinear force-displacement behavior) with different yield strengths. All structures had 5% viscous damping. The computed response for each structure is shown in Figures 6-9.

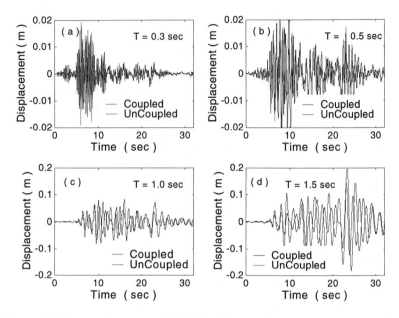

Figure 5. Computed response of SDOF oscillator to identical rock motions using coupled and uncoupled procedures: (a) $T = 0.3$ sec, (b) $T = 0.5$ sec, (c) $T = 1.0$ sec, and (d) $T = 1.5$ sec.

Table 1. Parameters used in analyses with linear and nonlinear structural response. Parameters k_1 and k_2 represent pre- and post-yield stiffnesses, respectively, of SDOF structure.

Parameters	$T = 0.5$ sec			$T = 1.0$ sec		
	Case 1	Case 2	Case 3	Case 1	Case 2	Case 3
W (kN)	1000	1000	1000	1000	1000	1000
k_1 (kN/m)	16102	16102	16102	4026	4026	4026
k_2 (kN/m)	16102	1610	1610	4026	403	403
f_y (kN)	200	200	100	200	200	100

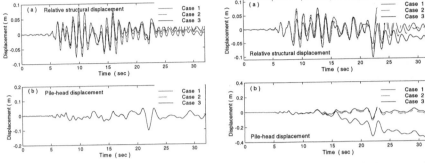

Figure 6. Time histories of (a) structural displacement and (b) pile head displacement for piles supporting linear and nonlinear structures with $T = 0.5$ sec.

Figure 7. Time histories of (a) structural displacement and (b) pile head displacement for piles supporting linear and nonlinear structures with $T = 1.0$ sec.

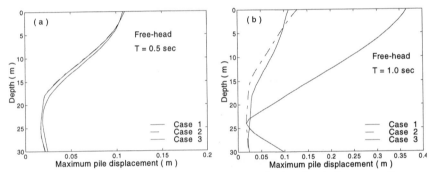

Figure 8. Profiles of maximum pile displacement for free-head piles supporting linear and nonlinear structures with (a) $T = 0.5$ sec, and (b) $T = 1.0$ sec.

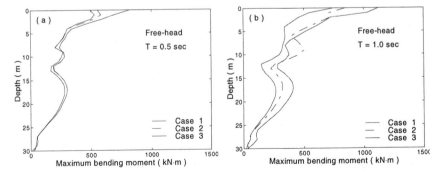

Figure 9. Profiles of maximum pile bending moment for free-head piles supporting linear and nonlinear structures with (a) $T = 0.5$ sec, and (b) $T = 1.0$ sec.

The results of these analyses indicate that yielding of the superstructure may not strongly influence the response of piles with fixed-head boundary conditions, but that the superstructure's influence on the pile with free-head boundary conditions can be significant. Structural yielding can strongly affect the structural displacements, generally, but not always, increasing them. The pile displacements may not be affected much by the structure yielding, especially for the fixed-head cases. Permanent relative structure displacements remain at the end of the earthquake shaking.

Decrease of the yielding force of the structure may or may not increase the displacements of the structures and piles. Generally, structural yielding decreases pile bending moments, especially for piles with free-head boundary conditions. The lower the yielding force, the more the pile bending moment decreases. This suggests that considering the nonlinearity of structures may be beneficial from the standpoint of pile demands.

Pile Stiffness Modeling

The capabilities of WAVE and DYNOPILE were used to make improved estimates of the stiffnesses of single piles in liquefiable soils. A procedure was developed to compute pile stiffnesses and express them in chart form similar to that already used in the *Manual*.

Pile stiffness can be expressed in different ways, particularly for a problem as complicated as the stiffness of a pile in a liquefying soil deposit. In the field, the stiffness of the pile will change with time as the inelastic response develops in the soil and as the soil softens and weakens due to excess porewater pressure generation. Since WSDOT had a strong interest in the displacements of pile foundations under earthquake loading, the stiffness was defined as the ratio of the static load (horizontal load or moment) applied to the head of the pile to the maximum displacement (or rotation) of the head of the pile. The procedure used to generate the pile head stiffness charts was as follows:

1. Obtain the free-field response for each selected site and loading condition using WAVE.
2. Perform DYNOPILE analyses with and without static forces (horizontal force or moment) applied at the top of the pile.
3. Compute the time history of relative deformation by subtracting the pile head deformation (horizontal displacement or rotation) time history for the case without an external load from the time history for the case with an external load. Select the maximum absolute value from the subtracted time history as the dynamic pile head deflection corresponding to that load level.
4. Compute the single pile stiffness as the ratio of the static load to the dynamic pile head deflection.

5. Perform a series of analyses with different static loading levels to generate a load-deflection curve for that pile type, soil site condition, and input motion.

The results of these analyses are illustrated in terms of load-displacement diagrams for free- and fixed-head piles subjected to static horizontal pile head loads in Figure 10, and in terms of moment rotation diagrams for free-head piles subjected to static overturning moment in Figure 11. As illustrated in both figures, the occurrence of liquefaction under dynamic loading conditions causes pile head displacements and/or rotations to increase for a given static load and/or moment. The load-displacement and moment-rotation behavior exhibited under static loading conditions is rather ductile; after an initial portion of nearly linear response, the resistance continues to increase with increasing deformation under static conditions. Under dynamic loading conditions, however, a peak resistance to horizontal load is reached shortly after initial yielding and deformations and nearly perfectly plastic behavior are exhibited thereafter.

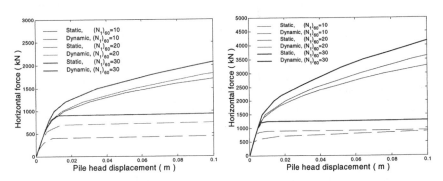

Figure 10. Pile head displacements for different static horizontal forces under static and dynamic loading conditions: (a) free-head, and (b) fixed-head conditions.

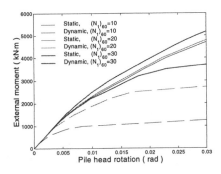

Figure 11. Pile head rotations for different static overturning moments under static and dynamic loading conditions.

Table 6.2 gives the initial pile stiffness values (dynamic) for foundation type P4 for three different soil densities. It also lists the stiffness values for a single pile of type P4 given in the *Manual*. It can be seen that the initial static stiffnesses are very close, in numerical values, to the initial dynamic stiffnesses. The dynamic initial stiffness values are slightly higher for denser soils than for looser soils. Compared with the values in the *Manual*, however, the initial stiffnesses obtained using DYNOPILE are higher than the corresponding stiffness values presented in the *Manual*.

Table 2. Initial dynamic stiffnesses for pile P4

Stiffness	$(N_1)_{60} = 10$	$(N_1)_{60} = 20$	$(N_1)_{60} = 30$	Manual
K_h (free-head) (kN/m)	1.134 x 10^5	1.137 x 10^5	1.139 x 10^5	0.438 x 10^5
K_h (fixed-head) (kN/m)	2.235 x 10^5	2.243 x 10^5	2.257 x 10^5	1.146 x 10^5
K_θ(free-head) (kN-m/rad)	2.716 x 10^5	2.847 x 10^5	2.8851 x 10^5	-

Figures 12 and 13 show the normalized stiffness reduction curves for translational (free-head and fixed-head) and rotational stiffness, respectively. The translational stiffness reduction curves for P4 from the *Manual* are also plotted in Figure 12. At small horizontal displacement levels (e.g. less than 0.01 m), the curves given by the *Manual* fit quite well with the curves derived from the DYNOPILE analyses. But at larger displacements, the reduction curves from DYNOPILE analyses drop much faster than the curves from the *Manual*. For example, when the horizontal displacement is larger than 0.05 m, the translational stiffness, predicted by DYNOPILE, is only about 10% of the initial value, while in the case of the *Manual*, it is about 30% of the initial value. These differences reflect the more accurate modeling of liquefaction and its effects on soil-pile interaction that was used in this investigation. This shows that, although the initial stiffness values given by the DYNOPILE are higher than those given in the *Manual*, DYNOPILE predicts lower stiffness values in the normal working displacement range.

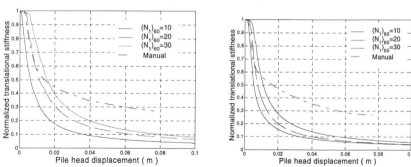

Figure 12. Normalized pile head stiffnesses for P4 and different soil densities: a)free-head conditions, b) fixed-head conditions

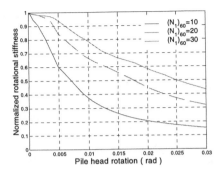

Figure 13. Normalized pile head rotational stiffnesses for P4 and different soil densities.

SUMMARY AND CONCLUSIONS

The purpose of the research described in this paper was to develop tools and procedures for evaluation of the stiffness of pile foundations in liquefiable soils during earthquakes. These tools and procedures were to be based on up-to-date models for liquefiable soil and for soil-pile interaction, which would obviate the need for many of the simplifying assumptions commonly used in practice.

For this purpose, a greatly improved model for description of the seismic response of liquefiable soil was implemented into a nonlinear, effective stress site response analysis (WAVE). Soil-pile interaction analyses were performed using an extended version of the program, DYNOPILE. DYNOPILE was modified to allow different pile head loading conditions, including the attachment of a single-degree-of-freedom structure to the pile head to allow coupled analysis of soil-pile-structure interaction. The effect of pore pressure generation on pile response as well as the effect of coupled analysis and nonlinear response of the structure was demonstrated.

The modified versions of the WAVE and DYNOPILE programs were used to improve and extend the stiffness charts for liquefiable soils that were presented in the *Design Manual for Foundation Stiffness under Seismic Loading* increasing both their accuracy and range of applicability.

REFERENCES

GeoSpectra, (1997), Design Manual for Foundation Stiffnesses under Seismic Loadings, Washington Department of Transportation.

Hamada, M. (1992). "Large Ground Deformations and their Effects on Lifelines: 1964 Niigata Earthquake," Case Studies of Liquefaction and Lifeline Performance During Past Earthquake Approaches, Technical Report NCEER-92-0001,

National Center for Earthquake Engineering Research, Buffalo, New York, Vol. 1, Sec. 3, pp. 1-123.

Horne, J. C. (1996), Effects of Liquefaction-Induced Lateral Spreading on Pile Foundations, Ph.D. Dissertation, University of Washington.

Horne J. C. and Kramer, S. L. 1998, Effects of Liquefaction on Pile Foundations, Research Report, Transportation Research Center, Research Project T9903, Task 28.

Kramer, S. L., Arduino, P., Li, P., and Baska, D. (2002), Dynamic Stiffness of Piles in Liquefiable Soils, Research Report, WA Transportation Research Center, Report Project T9903, Task A4 – Report No WA-RD 514.1.

Lysmer, J., Tabatabaie, M., Tajirian, F., Vahdani, S., and Ostadan, F. (1981). SASSI – A system for seismic analysis of soil-structure interaction, Report Number UCB/GT/81-02, Geotechnical Engineering, Department of Civil Engineering, University of California, Berkeley.

Nogami, T., Otani, J., Konagai, K., and Chen, H. L. (1992), Nonlinear Soil-Pile Interaction Model for Dynamic Lateral Motion, Journal of Geotechnical Engineering, ASCE, 118(1), pp. 89–116.

Reese, L. C. and Sullivan, W. R., (1980), Documentation of Computer Program COM624, Geotechnical Engineering Center, University of Texas at Austin.

LATERAL SUBGRADE MODULI FOR LIQUEFIED SAND
UNDER CYCLIC LOADING

Travis M. Gerber, Member, ASCE[1], Kyle M. Rollins, Member, ASCE[2]

ABSTRACT

The load-displacement response of liquefied soils surrounding deep foundations has been investigated previously by conducting lateral load tests on a full-scale 3 x 3 steel pipe pile group embedded in sands liquefied by controlled blasting. Based on this testing, p-y curves for fully liquefied sand were developed which included stiffness degradation due to multiple large-displacement load cycles. Additional study of the soil resistance-displacement time histories used to develop these p-y curves indicates that p-y curves for liquefied sand exhibit a generally linear response during an initial, relatively large-displacement cycle of loading. As the number of loading cycles increases, the stiffness of the soil response decreases, and p-y curves become increasingly shaped concave-up. The shapes of the p-y curves during initial and subsequent loading cycles are markedly different than that of a p-y curve for soft clay in the presence of water. One reason for this difference is that the soft clay model cannot reproduce the effects of contractive/dilative phase transformations. The lateral subgrade modulus for liquefied soil decreases rapidly with the first few loading cycles, losing between 50 to 70% of its initial value by the third cycle for a wide range of deflection levels.

INTRODUCTION

The lateral load-displacement response of soils surrounding deep foundations can be defined in terms of a lateral subgrade modulus. If the response is non-linear, the modulus is typically expressed in the form of a p-y curve. While p-y curves have been developed for several types of soil, p-y curves representative of liquefied sands

---

[1] Assist. Prof., Dept. of Civil & Environ. Eng., Brigham Young Univ., Provo, Utah.
[2] Prof., Dept. of Civil & Environ. Eng., Brigham Young Univ., Provo, Utah.

is a focus of on-going research. The availability of p-y curves for liquefied sand would permit the use of traditional beam-on-elastic-foundation (i.e. Winkler beam/spring) models to estimate the lateral resistance of deep foundations situated in liquefied soil under seismic loads.

To better define the lateral load-displacement response of liquefied soils, full-scale testing of several deep foundation systems was performed at the National Geotechnical Experiment Site on Treasure Island in California. To simulate the effects of seismically induced ground liquefaction, controlled blasting within the subsurface was used to produce elevated pore water pressures. Details of the Treasure Island Liquefaction Test (TILT) program are provided by Ashford and Rollins (2002). Using test data from the TILT program, p-y curves for fully liquefied sand were developed for a 3 x 3 pile group consisting of 324 mm O.D. steel pipe piles (Rollins et al., 2005) and a 0.6 m cast-in-steel-shell (CISS) pile (Weaver et al., 2005). One potential shortcoming of the developed p-y curves is that they represent the load-displacement response of liquefied soils only after the application of multiple large (~ 200 mm displacement) load cycles. Hence, the p-y curves presented in Rollins et al. (2005) and Weaver et al. (2005) contain cyclic loading effects and, consequently, represent a lower bound range of soil response. This paper presents results from subsequent study of the steel pipe pile p-y curves which describe how p-y curves for liquefied sand vary as a function of successive load cycles.

SITE CONDITIONS

Treasure Island is a 162 ha man-made island located in San Francisco Bay, California. The island consists of hydraulically-placed fill dredged from the bay. The loose fine sands and silty sands that constitute the uppermost portion of the soil profile are susceptible to liquefaction, with sand boils and ground deformations occurring during the 1989 Loma Prieta earthquake (Faris and Alba, 2000).

Site-specific subsurface explorations were conducted at TILT program test locations. Figure 1 presents the subsurface stratigraphy for the 3 x 3 pile group test site, together with the results of cone penetration tests (CPT) and standard penetration tests (SPT). Based on the CPT and SPT results, the relative densities of the soils are typically about 50% for the upper sands and 40% in the deeper, siltier sands. These materials are susceptible to liquefaction.

EXPERIMENTAL METHODS AND DATA

The 3 x 3 pile group consisted of 324 mm O.D. pipe piles made of ASTM A 252 Grade 3 steel, with a standard wall thickness of 9.5 mm. To protect externally mounted strain gauges, angle irons made of lower grade steel were welded on opposing sides of each pile. The piles were driven open ended at a spacing of 3.3-pile diameters, to a depth of approximately 11.5 m. Piles were connected together by a load frame with pivoting connections to create free-headed pile conditions. Displacement of the pile group was achieved by using a 2.2 MN hydraulic load

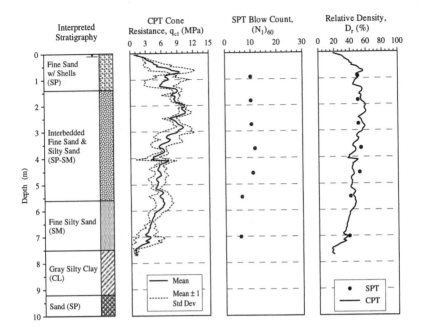

Figure 1: Subsurface stratigraphy and exploration data for 3 x 3 pile group test site
(after Ashford and Rollins, 2002)

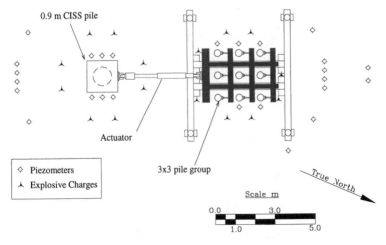

Figure 2: 3 x 3 pile group test setup (after Rollins et al., 2005)

actuator with controlled displacement reacting against an adjacent 0.9 m diameter CISS pile. The locations of the various test elements are shown in Figure 2.

Instrumentation for the test consisted of load cells and string potentiometers located on the piles and load frame, and strain gauges distributed along the length of the piles. Pore pressure transducers (PPT) were installed at various locations within the subsurface to monitor pore water pressures during the tests. Explosive charges used to induce liquefaction were located circumferentially about the pile group and CISS pile. The locations of the pore pressure transducers and explosive charges are shown in Figure 2.

The explosive charges appeared to be very effective in liquefying the sandy materials down to a depth of at least 5 m, which was the depth of the lowermost PPT. (Subsequent analysis suggests that pore pressures were significantly elevated to a depth of ~7.5 m). In terms of an excess pore pressure ratio, R_u, (being defined as the measured excess pore pressure divided by the initial vertical effective stress, with a value of 1.0 or 100% indicating complete liquefaction), the average values were in excess of 95% at most depths during the first 12 minutes of load testing. Figure 3 presents the time history of the excess pore pressure ratio measured by PPT 57, which is located within the perimeter of the 3 x 3 pile group at a depth of 1.98 m. Complete time histories for the excess pore pressure ratios can be found in Rollins et al. (2005).

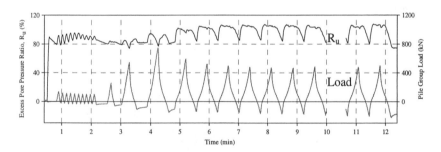

Figure 3: Time history of excess pore pressure ratio measured by PPT 57

The load testing which immediately followed the blasting consisted of the following series of load cycles: ten smaller 40 mm pushes, and then one 75 mm push, one 150 mm push, and eleven 230 mm pushes. Subsequent load cycles were applied intermittently during the course of the following hour. When referring to the load cycles, the convention used in the TILT program was to cite the nominal maximum displacement occurring at the load point during the cycle (e.g. the second 230 mm load cycle). Since the test was self-reacting, the maximum displacement was experienced by the weaker of the two foundation systems, which in this case was the 0.9 m CISS pile. Consequently, the maximum displacement at the pile load point within the 3 x 3 group was less than 200 mm during the "230 mm" load cycle. A continuous load-displacement curve for the pile group immediately following the blasting is shown in Figure 4. The load-displacement curve for a load cycle completed in the opposite direction before the blasting is shown for reference.

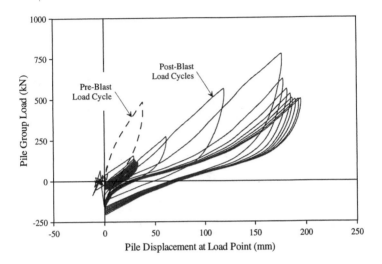

Figure 4: Load-displacement curves for 3 x 3 pile group

RESULTS

P-Y Curves Developed for Liquefied Sand

Curvature in the piles was derived from strain gauge measurements at discrete locations along the pile. Then, with these profiles of pile curvature, the curvature-area method from beam mechanics was used to determine the displacement of the pile with depth. The resulting deflection profiles were compared with the measured displacement of the pile heads, and the agreement between the measured and calculated values was found to be very good. To determine the soil pressure profile along the piles, bending moment profiles were first developed from a curvature-moment relationship derived from a section analysis of the pile cross-section. Next, a series of cubic polynomials were fit through successive sets of moment data points, and then these polynomials were twice differentiated. A comparison of the area under the calculated soil pressure diagram to the measured pile head load revealed generally good agreement. The resulting soil pressure- deflection pairs for each time increment were plotted to form continuous p-y curves for the entire load history. Representative p-y curves for the depths at which strain gauges were located are shown in Figure 5. Also shown in Figure 5 are the average excess pore pressure ratios, R_u, for each depth during the 12 minutes of testing represented by the p-y curves. At many depths, the soil was fully liquefied (as indicated by a $R_u \approx 100\%$). Details regarding the complete process used to develop the p-y curves for liquefied sand presented by Rollins et al. (2005) are described by Gerber (2003).

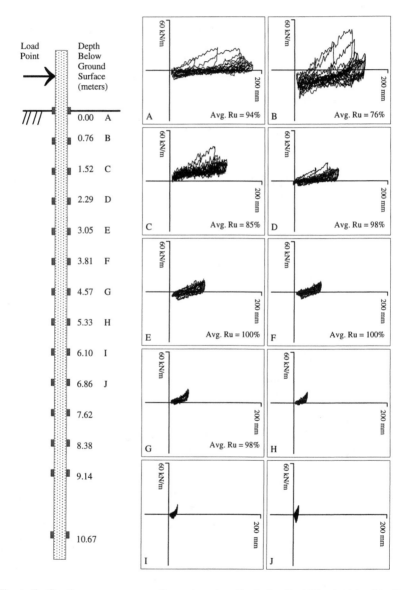

Figure 5: Continuous p-y curves for east center pile during first 12 minutes of testing

In order to validate the p-y curves developed for liquefied sand, representative segments of the continuous p-y curves were isolated and smoothed to form discrete p-y curves, and then these discrete curves were used in push-over analyses. These p-y curves for liquefied sand are shown in Figure 6. Since the continuous p-y curves began to approach a steady-state response after a few initial, larger load cycles (and since this response was a lower bound response), the discrete p-y curves (which were intended to be representative, and yet "conservative") were selected from the latter portion of the loading time history. Hence, the p-y curves shown in Figure 6 represent soil responses inclusive of cyclic loading and corresponding stiffness degradation. A comparison of pile head displacement and load, and pile bending moments calculated in the push-over analyses using the discrete p-y curves yielded a good agreement with measured values over a wide range of pile head loading (Rollins et al., 2005).

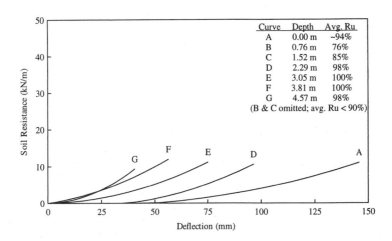

Figure 6: Discrete p-y curves for liquefied sand

Degradation of P-Y Curves during Cyclic Loading

In order to help quantify the effects of cyclic loading inherent in the p-y curves for liquefied sand shown in Figure 6, the continuous p-y curve at a depth of 2.29 meters (average R_u = 98%) was examined in detail. An enlarged view of this p-y curve is presented in Figure 7. In Figure 8, p-y curves representing the individual load paths from the start of each loading cycle are shown, with the data having been smoothed to eliminate irregularities due to both noise in the original strain gauge data and the double differentiation procedure used to determine the soil pressure. The p-y curves exhibit a small initial offset due to pile drift which occurred during the previous loading sequence consisting of ten 40 mm load cycles.

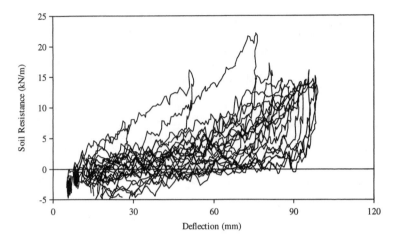

Figure 7: Continuous p-y curve at a depth of 2.29 m with average $R_u = 98\%$

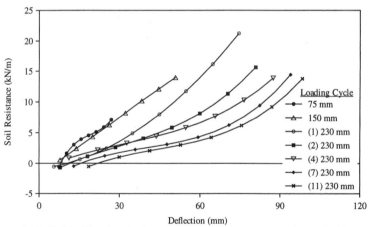

Figure 8: Individual load paths from the continuous p-y curve shown in Figure 7,
beginning at the start of each loading cycle (depth = 2.29 m)

As shown in Figure 8, the initial portion of the p-y curve for 75 mm loading cycle
has a concave-down shape, up to a deflection of 20 mm. Prior to this point in time
(~ 2.6 minutes), the time history presented previously in Figure 3 shows the excess
pore pressure ratio increasing with increasing load (and deflection). Immediately
beyond 20 mm, however, the excess pore pressure ratio decreases slightly as both
load and deflection increase. This behavior suggests that at this point in time, the soil
is transitioning from contractive to dilative behavior. The subsequent 150 mm

loading cycle is fairly linear. As deflection exceeds 20 mm, the load path coincides with the load path from the previous 75 mm load cycle. As the load path exceeds the maximum previous deflection, the excess pore pressure ratio drops markedly, indicating another phase transformation. The next loading cycle – the first 230 mm push – exhibits a concave-up shape and is generally flatter than the previous loading cycle. The excess pore pressure ratio again exhibits a significant drop as the load path exceeds the maximum previous deflection. The next few 230 mm loading cycles exhibit both decreasing stiffness and more pronounced concave-upward curvature. The last few loading cycles exhibit nearly steady-state load-response.

In order to quantify the degradation of stiffness during cyclic loading, it was necessary to define equivalent loading cycles since the first two loading cycles had smaller and differing ranges of displacement than subsequent loading cycles. For example, at a displacement of 10 mm, the first 230 mm loading cycle would represent the third loading whereas at a displacement of 60 mm, the same first 230 mm loading cycle would represent the first loading. Data from the individual load paths shown in Figure 8 was assessed according to specific displacement levels, the number of times those displacement levels had been reached or exceeded, and the associated soil resistances. With the aid of statistical regressions, the p-y curves presented in Figure 9 were developed. These p-y curves are quite similar to those shown in Figure 8, except that p-y curves have been shifted by a constant value so that they coincide with the origin. The p-y curves shown in Figure 9 are believed to approximate what the p-y curves for liquefied sand would have been at a depth of 2.29 m had all the loading cycles been similarly large.

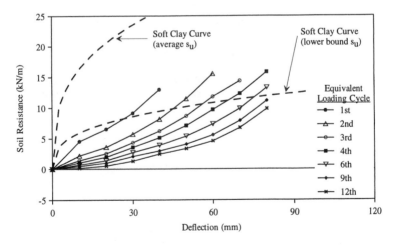

Figure 9: P-y curves for liquefied sand as a function of deflection level and loading cycle (depth = 2.29 m)

Inspection of Figure 9 reveals that upon initial loading, the p-y curve appears to be relatively linear up to a displacement of 40 mm (except an elevated strength in the range of 10 mm). As the number of loading cycles increases, the stiffness begins to degrade, and the p-y curves begin to take on a more pronounced concave-up shape. Visual inspection of the continuous p-y curves presented in Figure 5 suggests that this characterization is generally applicable to p-y curves at other depths.

Also shown in Figure 9 are two p-y curves for soft clay in free water as defined by Matlock (1970). It has been suggested by others that p-y curves for soft-clay can be used to approximate p-y curves for liquefied sand. The upper p-y curve is based on a shear strength of 14.4 kPa and an ε_{50} value of 0.02. The lower p-y curve is based on a shear strength of 4.8 kPa and an ε_{50} value of 0.02. These two shear strengths represent the average and lower-bound residual undrained shear strengths for the 3 x 3 pile group test site at a depth of 2.29 m based on correlations with $(N_1)_{60}$ developed by Seed and Harder (1990). Neither soft clay p-y curve accurately describes any of the cyclic p-y curves for liquefied sand. The soft clay p-y curves are initially too stiff and do not possess the same shape over the entire range of deflection. The shapes should not be expected to match because, although the liquefied sand may be undrained, the sand undergoes a transformation from contractive to dilative behavior which increases strength with increasing deflection whereas the soft clay model does not account for dilatancy and is shaped increasingly concave-down as deflection increases.

The degradation of p-y curves for liquefied sand during cyclic loading is perhaps more readily quantified in terms of lateral subgrade modulus (i.e., soil resistance divided by deflection). Figure 10 shows how lateral subgrade modulus varies as a function of deflection level and number of loading cycles. Relative to the first loading cycle, the lateral subgrade modulus during the second loading cycle is reduced by 35 to 50%, depending upon the deflection level. For the third cycle, the reduction in modulus is 50 to 70% of its initial value. Hence, the stiffness of liquefied soil decreases dramatically after a few large displacement cycles. After approximately ten or so loading cycles, subsequent decreases in lateral subgrade modulus are essentially negligible.

Although detailed analysis of the p-y curves at other depths was not performed at this time, based on similarities between the continuous p-y curves at different depths, it is anticipated that the degradation of p-y curves at other depths are proportionally similar.

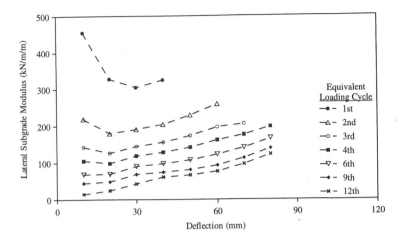

Figure 10: Lateral subgrade moduli for liquefied sand as a function of deflection
level and loading cycle (depth = 2.29 m)

CONCLUSIONS

Based on the data, analysis, and results presented in this paper, the following
conclusions are made:

1. P-y curves for liquefied sand based on full-scale testing of a deep foundation
   exhibit a generally linear response during an initial, relatively large-
   displacement cycle of loading.

2. As the number of loading cycles increases, the stiffness of the soil response
   decreases, and the p-y curves become increasingly shaped concave-up at the
   upper range of displacement.

3. The shapes of the p-y curves during initial and subsequent loading cycles are
   markedly different than that of a p-y curve for soft clay in the presence of water.
   One reason for this difference is that the soft clay model cannot reproduce the
   effects of contractive/dilative phase transformations.

4. The lateral subgrade modulus for liquefied soil decreases rapidly with the first
   few loading cycles, losing between 50 to 70% of its initial value by the third
   cycle for a wide range of deflection levels.

REFERENCES

Ashford, S. A. and Rollins, K. M. (2002). *TILT: the Treasure Island Liquefaction
Test, Report Number SSRP 2001/17*. Dept. of Structural Engineering, University
of California at San Diego, California.

Faris, J. R., and de Alba, P. (2000). "National geotechnical experimentation site at
Treasure Island, California." *National Geotechnical Experimentation Sites,*

Geotechnical Special Publication No. 93, Benoit, J., and Lutenegger, A. J., eds., ASCE, New York, 52-71.

Gerber, T. M. (2003). "P-y curves for liquefied sand subject to cyclic loading based on testing of full-scale deep foundations." PhD dissertation, Brigham Young University, Provo, Utah.

Matlock, H. (1970). "Correlations for design of laterally loaded piles in soft clay." *Proc., 2nd Offshore Technology Conf., OTC 1204*. Vol. 1, Houston, 577-594.

Rollins, K. M., Gerber, T. M., Lane, J. D., and Ashford, S. A. (2005). "Lateral resistance of a full-scale pile group in liquefied sand." *J. of Geotechnical and Geoenvironmental Engineering*, 131(1), 115-125.

Seed, R. B., and Harder, L. F. (1990). "SPT-based analysis of cyclic pore pressure generation on undrained residual strength." *Proc., H. Bolton Seed Memorial Symposium*, University of California at Berkeley, California, 351-376.

Weaver, T. J., Ashford, S. A., and Rollins, K. M. (2005). "Response of 0.6 m cast-in-steel-shell pile in liquefied soil under lateral loading." *J. of Geotechnical and Geoenvironmental Engineering*, 131(1), 94-102.

OBSERVATIONS AND ANALYSIS OF PILE GROUPS IN LIQUEFIED AND LATERALLY SPREADING GROUND IN CENTRIFUGE TESTS

Scott J. Brandenberg, Assoc. Member, ASCE[1], Ross W. Boulanger, Member, ASCE[2], Bruce L. Kutter, Member, ASCE[2], Dongdong Chang, Student Member, ASCE[3]

ABSTRACT

Static seismic analysis procedures are evaluated against results from eight dynamic model tests performed on a 9-m-radius centrifuge to study the behavior of pile groups in liquefiable and laterally spreading ground. Either 0.73-m-diameter piles or 1.17-m-diameter piles were fixed into a large embedded pile cap in each test. The soil profile consisted of a gently sloping nonliquefied crust over liquefiable loose sand over dense sand. Each model was tested with a series of realistic earthquake motions with peak base accelerations ranging from 0.13 g to 1.00 g. Representative data that characterize important aspects of soil-pile interaction mechanics in liquefied and laterally spreading ground are presented. The observed results of static seismic analyses using Beam on Nonlinear Winkler Foundation (BNWF) models are compared with the centrifuge test data. The analyses are shown to reasonably predict peak bending moments and pile cap displacements when input parameters correspond well with test measurements, but significant errors arise when input parameters do not match the test observations even though they do correspond with commonly-used design assumptions.

---

[1] Graduate Student, Univ. of California, Davis, Dept. of Civil and Environmental Engineering, Davis, CA 95616, E-mail: sjbrandenberg@ucdavis.edu.
[2] Professor, Univ. of California, Davis, Dept. of Civil and Environmental Engineering, Davis, CA 95616.
[3] Graduate Student, Univ. of California, Davis, Dept. of Civil and Environmental Engineering, Davis, CA 95616.

INTRODUCTION

Loads from laterally spreading ground have been a major cause of damages to pile foundations in past earthquakes. Analyses of case histories have shown that damages are particularly intense when a nonliquefied surface crust layer spreads laterally over underlying liquefied layers. The complex dynamic response of pile foundations during earthquakes is not directly accounted for in static seismic analysis procedures, but is rather represented through the specification of approximately equivalent static loading conditions. Quantifying the uncertainty in static seismic analysis procedures and their sensitivity to the specified loading conditions and modeling parameters requires comparisons against a range of case history or physical model test data.

This paper presents an evaluation of a static seismic analysis procedure using results from a series of eight dynamic centrifuge model tests on a 9-m-radius centrifuge to study the behavior of pile foundations in laterally spreading ground. The soil profile in each test consisted of a gently-sloping nonliquefiable crust over liquefiable loose sand over dense sand. Seven of the models contained a six-pile group with a large pile cap embedded in the nonliquefiable crust. A series of realistic earthquake motions was applied to the base of each of the models. Representative data recordings are presented, as are soil-structure interaction forces obtained by processing the raw recorded data. Static seismic analyses of the pile groups were performed for the entire set of centrifuge test data to determine the input parameters and specified equivalent static loading conditions that most significantly influence analysis results. Recommendations for design are derived from the centrifuge test observations and results of the suite of analyses.

CENTRIFUGE TESTS AND OBSERVATIONS

Eight dynamic centrifuge tests were performed on the 9-m radius centrifuge at the Center for Geotechnical Modeling at the University of California, Davis to study the behavior of pile foundations in liquefied and laterally spreading ground. Models were tested in a flexible shear beam container (FSB2) at centrifugal accelerations ranging from 36.2 g to 57.2 g. Results are presented in prototype units. Complete data reports from the centrifuge tests are presented by Singh et al. (2000a,b, 2001), and Brandenberg et al. (2001a,b, 2003). The data are available on the Center for Geotechnical Modeling website (http://nees.ucdavis.edu). A small portion of the data is presented in this paper.

Soil Properties

Fig. 1 shows the model layout for centrifuge test SJB03, which was similar to six other tests containing a six-pile group with a large embedded pile cap. The soil profile for all of the models consisted of a nonliquefiable crust overlying loose sand ($D_r \approx 21\text{-}35\%$) overlying dense sand ($D_r \approx 69\text{-}83\%$). All of the layers sloped gently toward a river channel carved in the crust at one end of the model. The nonliquefiable crust consisted of reconstituted Bay mud (liquid limit $\approx$ 88, plasticity index $\approx$ 48) that was mechanically consolidated with a large hydraulic press, and subsequently carved to the desired slope. The sand layers beneath the crust consisted of uniformly-graded Nevada Sand (C_u = 1.5, D_{50} = 0.15 mm). A thin layer of coarse Monterey sand was placed on the surface of the Bay mud for some of the models. Water was used as a pore fluid for all of the models.

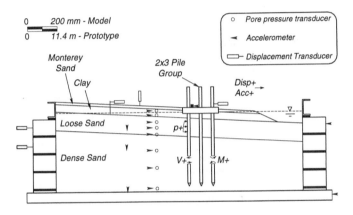

Figure 1: Sketch of centrifuge model SJB03. Much of the dense array of nearly 100 instruments, including 40 Wheatstone strain gauge full bridges on the piles, has been excluded from the figure for clarity.

Foundation Properties

The seven most recent model tests each contained a six-pile group with a large pile cap embedded in the nonliquefied crust layer. Two different pile diameters were tested (0.73 m and 1.17 m) and the center-to-center spacing of the piles was four diameters. The pile caps for the six-pile groups provided a stiff rotational restraint at the connection between the piles and the pile caps with the measured rotational stiffness being about 1,300,000 kN·m/rad for the tests at 57.2 g and 390,000 kN·m/rad for the tests at 38.1 g. Single-degree-of-freedom structures with fixed-base natural periods of 0.8 s and 0.3 s were connected to the pile cap for tests DDC01 and DDC02, respectively. The first test contained three single piles of various diameters and a two-pile group, and detailed analysis of the piles was presented by Boulanger et al. (2003).

Simulated Earthquakes

Each test was shaken with a number of simulated earthquakes conducted in series with sufficient time between shakes to allow dissipation of excess pore pressures. The simulated earthquakes were scaled versions of the acceleration recordings either from Port Island (83-m depth, north-south direction) during the Kobe earthquake, or from the University of California, Santa Cruz (UCSC/Lick Lab, Channel 1) during the Loma Prieta earthquake. These earthquake motions were chosen because they contain different frequency content and shaking characteristics. Generally, the shake sequence applied to the models was a small event ($a_{max,base}$ = 0.13g to 0.17g) followed by a medium event ($a_{max,base}$ = 0.30g to 0.45g) followed by one or more large events ($a_{max,base}$ = 0.67g to 1.00g).

Observations from Centrifuge Test Data

Several time series of raw and processed data are shown for the large Kobe motion for test SJB03 in Fig. 2 to illustrate the behavior of the soil and the pile group. The bending moment 2.7 m below the ground surface was recorded from the moment bridge on the southeast pile (SEM) that was closest to the pile cap connection, where the peak bending moments were measured during the test. Subgrade reaction loads, p, were calculated by double-differentiating distributions of bending moment (Boulanger et al., 2003; Brandenberg et al., 2005). Time series of p and r_u were located near the middle of the loose sand layer, and r_u was about 13 m down-slope of the pile group where the influence of the pile group is small. Sign conventions for displacement, acceleration, bending moment, shear and p are shown in Fig. 1. The large Kobe motion was the fourth in a series of four earthquake motions, therefore some of the values exhibit initial offsets representing prior seismic history.

At the time that the peak bending moment was measured during the large Kobe motion (-8840 kN·m), the following also occurred:

- The lateral load from the clay crust was 5730 kN, which was the maximum for the large Kobe motion.
- The pile cap inertia force was 5790 kN, which was the maximum for the test.
- The subgrade reaction, p, 6.7 m below the ground surface was −370 kN/m, which was a local minimum. Note that the subgrade reaction was negative; the loose sand restrained the pile from moving down-slope.
- The excess pore pressure ratio was 0.5, near a local minimum, in spite of having been close to 1.0 earlier in the shake.
- The total displacement of the pile cap (including 0.2 m of displacement induced by previous earthquakes) was 0.4 m, which was smaller than the peak cap displacement of about 0.5 m, but larger than the permanent pile cap displacement of about 0.3 m at the end of shaking.
- The displacement of the clay crust was 2.3 m, which was less than the permanent crust displacement of nearly 3.5 m.

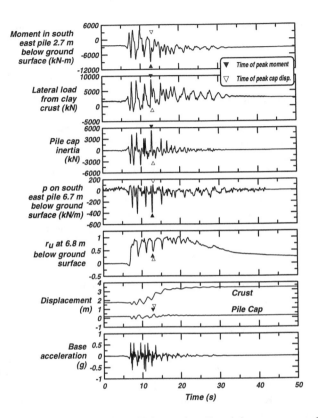

Figure 2: Time series from the large Kobe motion (fourth in a sequence of four motions) from test SJB03.

Several transient drops in pore pressure in the loose sand during shaking are important because they occurred at the same time as the peak bending moments and cap displacements, and are attributed to undrained shear loading of dilatant soil. Dilatancy is the tendency of sand to dilate during drained shear loading, which is manifested as in an increase in effective stress (decrease in pore pressure) during undrained shear loading. The sand temporarily becomes stiff and strong during the transient drops in pore pressure, which contrasts with the assumption that is commonly made in design practice that liquefied sand is soft and weak. Designers should recognize that static seismic analysis procedures, such as those presented later in this paper (and e.g. JRA, 2002; Dobry et al., 2003), often utilize crude approximations to envelope the critical loading patterns that occur under more complex dynamic loading conditions, and should use caution in extrapolating analytical models beyond the ranges of their experimental validation.

The following two observations from the test data are particularly important because they are contrary to potentially un-conservative assumptions commonly made in design:

1. Friction loads between the pile caps and laterally spreading crusts (back-calculated by Boulanger et al., 2003, Brandenberg et al., 2005) were nearly as large as the passive forces exerted on the upslope faces of the pile caps. Neglecting friction forces in design calculations based on the assumption that lateral loads are dominated by passive forces could be significantly unconservative.

2. Inertia loading from the pile caps (and any superstructures attached to the caps) acted in-phase with and in the same direction as the peak crust loads (Chang et al., 2005). Designers might decouple the kinematic and inertia loading in analysis based on the assumption that lateral crust loads act after strong shaking (when the ground surface displacements are largest), while peak inertia loads act during strong shaking (e.g. Martin et al., 2002), which could be significantly unconservative.

The effects of these two erroneous assumptions on the results of static seismic analyses of the centrifuge tests are presented later in this paper.

STATIC SEISMIC BNWF ANALYSIS METHOD AND RESULTS

Beam on Nonlinear Winkler Foundation (BNWF) analyses were performed for the six-pile groups tested in the centrifuge to determine how accurately such methods can predict peak bending moments and pile cap displacements, and to quantify potential errors that arise from alternative specifications for input parameters. Static seismic BNWF methods do not fundamentally capture complicated dynamic behavior such as dilatancy of liquefying sand, but they are commonly utilized in design practice. Comparisons between the BNWF analyses and the centrifuge test data can provide guidance in selecting reasonable analysis input parameters, which is critical for obtaining safe and reliable foundation designs.

The BNWF analysis method (Fig. 3) consisted of pile elements that were attached to nonlinear soil elements to represent soil-pile interaction. The properties of the nonlinear soil elements were softened to account for liquefaction of the sand layers, with the degree of softening depending on the sand's relative density (Boulanger et al., 2003). Horizontal displacements were imposed on the free ends of the p-y elements to represent lateral spreading free-field ground displacements. Forces were applied to the pile cap elements to represent the inertia loading imposed by the pile caps. The analyses were performed using the Open System for Earthquake Engineering Simulation (OpenSees) developed by the Pacific Earthquake Engineering Research (PEER) Center (http://opensees.berkeley.edu).

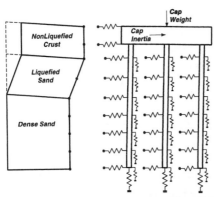

Figure 3: Schematic of BNWF analysis with imposed free-field soil displacements and inertia loads.

Results of a BNWF analysis of centrifuge model SJB03 for a large Kobe motion are presented in Fig. 4. The soil displacement profile, inertia loads and capacities of the p-y elements in the nonliquefied crust used in the analysis were based on parameters and guidelines that were calibrated with values measured during the tests. While a single set of parameters can result in an excellent fit with the data for any given test, the more robust design guidelines used in this study were calibrated with the entire suite of centrifuge tests and provide a measure of the uncertainty inherent in the assumptions. At the location of the top strain gauge bridge 2.7 m below the ground surface, the predicted bending moment was -8581 kN·m, which very reasonably matched the peak measured bending moment of -8840 kN·m (3% under-prediction of bending moment magnitude). The predicted pile cap displacement was 0.38 m, which is smaller than the peak total measured cap displacement of 0.5 m that was induced by all four earthquakes, but larger than the permanent cap displacement of about 0.3 m. While the peak values of bending moments and cap displacements were predicted reasonably well, the distributions of pile displacement and bending moment exhibited larger deviations from the measured distributions, but the deviations would likely be considered inconsequential for making practical design decisions.

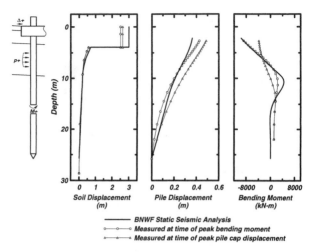

Figure 4: Results of a BNWF analysis for the large Kobe motion for test SJB03.

Predicted bending moments and pile cap displacements are plotted versus peak measured bending moments and pile cap displacements for each shake from five centrifuge tests in Fig. 5. The soil displacement profiles, inertia loads and capacities of the p-y elements in the nonliquefied crust were all estimated using design guidelines calibrated to the entire suite of test data. Hence, Fig. 5 represents the potential accuracy of the BNWF static seismic design approach when the input parameters closely match field conditions. Bending moments were predicted reasonably well with a small conservative bias, and cap displacements were also reasonable, but had an unconservative bias. The bias cannot be clearly attributed to any single source of error, and is related to the simplifications inherent in representing complex dynamic loading conditions using simplified static analysis methods. The bias in the cap displacement is larger at large displacement, which in part reflects the inability of the static analysis method to capture ratcheting of the foundation that occurs when plastic deformations progressively accumulate during repeated loading cycles.

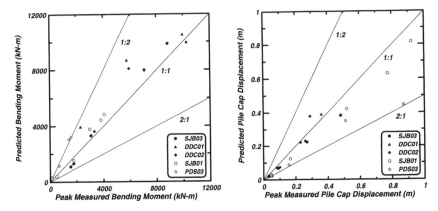

Figure 5: Predicted versus measured values of bending moment and pile cap displacement for BNWF analyses using soil displacement profiles, inertial loads and crust loads that were calibrated with the suite of test data. Solid symbols represent tests with 0.73-m diameter piles and open symbols represent tests with 1.2-m diameter piles.

Utilizing measured quantities as input parameters is insightful for demonstrating how well static seismic BNWF analyses can perform when reasonable input parameters are selected. However, it is unlikely that designers would select input parameters that so closely match field conditions due to inherent uncertainty, particularly considering the discrepancies between some of the observations from the centrifuge test data and the corresponding common design assumptions. For example, friction between the crust and the sides and base of the pile cap might be excluded from the computation of total load on the foundation based on the assumption that passive loading is the dominant source. Also, pile cap and structural inertia loads might be excluded from the analysis based on the assumption that peak crust loads and peak inertia loads do not act at the same time and can be uncoupled in analysis. The influence of these assumptions on the accuracy of the BNWF static seismic method is demonstrated in Fig. 6, in which friction loads and inertia loads have been omitted. The analyses in Fig. 6 were otherwise identical with those in Fig. 5. In both cases the prediction tends to be significantly un-conservative, with bending moments and cap displacements being under-predicted by an average factor of about 2.

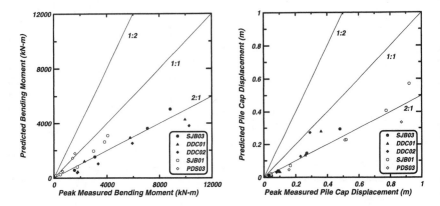

Figure 6: Predicted versus measured values of bending moment and pile cap displacement for BNWF analyses that neglect inertia forces and friction forces between the nonliquefied crust and the sides and base of the pile cap.

CONCLUSIONS

A series of eight dynamic centrifuge tests provided a detailed understanding of the complex dynamic loading mechanics that occur between pile foundations and laterally spreading soil. Some observed load transfer mechanics were contrary to common design assumptions. For example, large pile cap inertia forces were observed to act simultaneously with the peak crust loads, and friction forces between the sides and bases of the pile caps were shown to be nearly as large as the passive forces on the upslope faces of some of the pile caps.

A static seismic BNWF analysis procedure with associated guidelines for input parameter selection and equivalent static load representations produced reasonably accurate predictions of peak bending moments and pile cap displacements. The sensitivity of the overall predictions to alternative choices of input parameters and equivalent static loading conditions was evaluated. For example, neglecting inertia loading and friction loading between the nonliquefied crust and sides and base of the pile caps, both common design assumptions, resulted in BNWF predictions that were un-conservative by about a factor of two. Such loads should be included in analyses of pile groups that are designed to be stiff enough and strong enough to resist the lateral loads imposed by spreading nonliquefied crusts, similar to the pile foundations in the centrifuge tests.

Caution should be used in extrapolating test observations and design methods beyond their range of experimental validation. The pile group foundations in the centrifuge tests were sufficiently strong and stiff to mobilize the full lateral loading from the nonliquefied crust layer as the crust spread down-slope past the foundations.

Phasing between inertia loads and crust loads might be different for more flexible pile foundations that move along with the ground. Furthermore, the size of the laterally spreading soil mass in the centrifuge tests was sufficient to mobilize the full crust loading against all of the model pile group foundations. Therefore, applying "free-field" ground displacements in the BNWF analyses was appropriate. If the extent of the laterally spreading mass is small (i.e. finite-width bridge abutments), the pinning effect provided by the piles could limit the lateral spreading displacements such that the full lateral loads are not mobilized. In such cases, applying "free-field" ground displacements is inappropriate and consideration must be given to the influence of the foundation on the lateral spreading displacements (e.g. Martin et al., 2002).

ACKNOWLEDGMENTS

Funding was provided by Caltrans under contract numbers 59A0162 and 59A0392 and by the Pacific Earthquake Engineering Research (PEER) Center, through the Earthquake Engineering Research Centers Program of the National Science Foundation, under contract 2312001. The contents of this paper do not necessarily represent a policy of either agency or endorsement by the state or federal government. The centrifuge shaker was designed and constructed with support from the National Science Foundation (NSF), Obayashi Corp., Caltrans and the University of California. Recent upgrades have been funded by NSF award CMS-0086566 through the George E. Brown, Jr. Network for Earthquake Engineering Simulation (NEES). Center for Geotechnical Modeling (CGM) facility manager Dan Wilson, and CGM staff Tom Kohnke, Tom Coker and Chad Justice provided assistance with centrifuge modeling. Former UC Davis graduate student Priyanshu Singh oversaw some of the centrifuge tests, and performed some data processing and analysis of single piles.

REFERENCES

Boulanger, R. W., Kutter, B. L., Brandenberg, S. J., Singh, P., and Chang, D. (2003). Pile foundations in liquefied and laterally spreading ground during earthquakes: Centrifuge experiments and analyses. Report UCD/CGM-03/01, Center for Geotechnical Modeling, Univ. of California, Davis, CA.

Brandenberg, S.J., Boulanger, R.W., Kutter, B.L., and Chang, D. (2005). "Behavior of pile foundations in laterally spreading ground in centrifuge tests." Under review. *J. of Geotech. Geoenviron. Eng.*, ASCE.

Brandenberg, S. J., Chang, D., Boulanger, R. W., and Kutter, B. L. (2003). "Behavior of piles in laterally spreading ground during earthquakes – centrifuge data report for SJB03." Report No. UCD/CGMDR-03/03, Center for Geotechnical Modeling, Department of Civil Engineering, University of California, Davis.

Brandenberg, S. J., Singh, P., Boulanger, R. W., and Kutter, B. L. [2001 (a)]. "Behavior of piles in laterally spreading ground during earthquakes – centrifuge data report for SJB01." Report No. UCD/CGMDR-01/02, Center for Geotechnical Modeling, Department of Civil Engineering, University of California, Davis.

Brandenberg, S. J., Singh, P., Boulanger, R. W., and Kutter, B. L. [2001 (b)]. "Behavior of piles in laterally spreading ground during earthquakes – centrifuge data report for SJB02." Report No. UCD/CGMDR-01/06, Center for Geotechnical Modeling, Department of Civil Engineering, University of California, Davis.

Chang, D., Boulanger, R.W., Kutter, B.L, and Brandenberg, S.J. (2005). Accepted. "Experimental observations of inertial and lateral spreading loads on pile groups during earthquakes. " *Proc. ASCE Geo-Frontiers Conference*, Austin, TX, January 24-26, 2005.

Dobry, R., Abdoun, T., O'Rourke, T. D., and Goh, S. H. (2003). "Single Piles in Lateral Spreads: Field Bending Moment Evaluation." *J. Geotech. Geoenviron. Eng.*, ASCE, Vol. 129(10), 879-889.

JRA (2002). *Specifications for highway bridges.* Japan Road Association, Preliminary English Version, prepared by Public Works Research Institute (PWRI) and Civil Engineering Research Laboratory (CRL), Japan, November.

Martin, G. R., March, M. L., Anderson, D. G., Mayes, R. L., and Power, M. S. (2002). "Recommended design approach for liquefaction induced lateral spreads." Proc. Third National Seismic Conference and Workshop on Bridges and Highways, MCEER-02-SP04, Buffalo, N.Y.

Singh, P., Brandenberg, S. J., Boulanger, R. W., and Kutter, B. L. (2001). "Piles under earthquake loading – centrifuge data report for PDS03." Report No. UCD/CGMDR-01/01, Center for Geotechnical Modeling, Department of Civil Engineering, University of California, Davis.

Singh, P., Subramanian, P. K., Boulanger, R. W., and Kutter, B. L. [2000 (a)]. "Behavior of piles in laterally spreading ground – centrifuge data report for PDS01." Report No. UCD/CGMDR-00/05, Center for Geotechnical Modeling, Department of Civil Engineering, University of California, Davis.

Singh, P., Boulanger, R. W., and Kutter, B. L. [2000 (b)]. "Piles under earthquake loading – centrifuge data report for PDS02." Report No. UCD/CGMDR-00/06, Center for Geotechnical Modeling, Department of Civil Engineering, University of California, Davis

MODELING THE SEISMIC PERFORMANCE OF PILE FOUNDATIONS FOR PORT AND COASTAL INFRASTRUCTURE

Stephen E. Dickenson, Member, ASCE[1] and Nason J. McCullough, Assoc. Member, ASCE[2]

ABSTRACT

The utilization of performance-based seismic design concepts at major ports often necessitates a reliance on numerical models for simulating the dynamic soil-foundation-structure interaction of piers and wharves. Construction and siting of pile supported wharves, which comprise the most common type of waterfront structure in cargo handling portions of major U.S. ports, includes pile embedment in submarine slopes routinely comprised of rock-fill underlain by weak marine soils. Field observations at ports around the world lead to the general conclusion that incipient, permanent slope deformations can be expected at ground motion levels as low as 0.15g, well below design levels in many regions. In light of the resources required to prevent slope deformation small permanent slope deformations are considered acceptable for most waterfront projects. The allowance in seismic performance criteria of limited, yet permanent, deformations of the slope and piles makes the design of pile foundations for port waterfront structures rather unique. Seismic analysis requires that both the kinematic and inertial aspects of the seismic loading of pile foundations are evaluated in a coupled manner. This paper addresses the seismic analysis of pile supported wharves; however, aspects of this work can be readily applied to other coastal infrastructure.

In order to identify the strengths and limitations of practice-oriented methods for evaluating the seismic performance of pile supported wharves centrifuge modeling was employed to supplement field case history data for validation of two deformation-based analysis methods; a numerical 2D nonlinear, effective stress model, and a simple, straightforward procedure based on the rigid, sliding block concept. Modeling results

---

[1] Associate Professor, Oregon State University, Dept of Civil, Construction and Environmental Engineering, Corvallis, OR 97331.

[2] Geotechnical Engineer, CH2M HILL, Corvallis, OR 97330.

demonstrate that the practice-oriented, sliding block model can yield suitably accurat₁ deformation estimates, provided that the analysis method accounts for excess pore pressur₁ generation and the stabilizing effect of the piles in the slope. The numerical procedure, whicᒷ utilized a simplified stress-based pore pressure generation constitutive model, resulted i₁ reliable deformation patterns and acceleration estimates, and pore pressures were somewhaᵼ well reproduced; however the maximum dynamic and residual moments in the piles were nᴏ well replicated. Shortcomings of the numerical geomechanical modeling are addressed. Th₁ limitations notwithstanding, the numerical procedure is advantageous compared to the slidin₁ block method, as the pattern of soil deformation and pile curvature are more accurateᒷ simulated, a more complete picture of soil-foundation-structure interaction is provided, an₁ the coupled performance of the waterfront slope and wharf is more clearly demonstrated.

INTRODUCTION

Pile-supported wharves, as discussed herein, refer to the *system* of piles, the waterfroᴎ slope (fill and/or native soil), backfill soils, and the wharf deck itself. In some instances th₁ construction of wharves employ a cut-slope in which existing soil is excavated; however, th₁ majority of pile-supported wharf construction involves reclaimed land utilizing a perimete₁ dike to retain backfill soil. The configuration of the containment dikes, slope, pil₁ foundation, and other wharf characteristics will reflect port-specific site conditions and carg₁ handling requirements. Two common examples, as replicated in the centrifuge, are illustrate₁ in Figure 1. Although these configurations may seem similar in overall appearance th₁ differences in the dike or rock-fill geometry may result in dissimilar seismic performance. This is especially relevant for the seismic performance of pile foundations that exten₁ through the fill layers and native foundation soils. Variations in the seismic response of th₁ layered rock-fill, sand backfill, and foundation soils lead to localized patterns of groun₁ deformation, with associated concentration of stresses in piles. The patterns of groun₁ deformation and anticipated performance of the piles will likely vary in the two cases showr despite equivalent computed factor of safety against slope deformation obtained from limi₁ equilibrium methods.

Field case histories demonstrate that damage to pile supported wharves is directly relate₁ to the magnitude of permanent ground deformations (Werner 1998, PIANC 2001). It i₅ pertinent to note that in most of the documented cases wharf damage (ranging from minᴑ repairable damage to failure) has occurred at ground motion levels that were much smalle₁ than current design motions at the respective ports. These observations combined with th₁ performance-based seismic design requirements adopted at many ports highlight the need fo₁ analysis methods capable of reliably predicting permanent ground deformations in orde₁ assess the kinematic loading of piles. Two methods are often utilized in practice for th₁ estimation of permanent ground deformations:

1) Newmark-type, rigid, sliding block procedures. This well-known procedure is base₁ on limit-equilibrium methods modified to include the effect of soil degradation (i.e. excess pore pressure generation and strain softening) and the beneficial effect of piles in the slope.
2) Numerical procedures incorporating coupled non-linear, effective stress models. These models are capable of representing the full soil continuum and simulating soil-pile interaction in a coupled manner. In addition, they are often capable of modeling

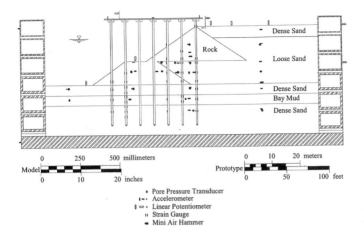

(a)   Wharf supported on a multi-lift rock fill dike (Centrifuge model NJM02).

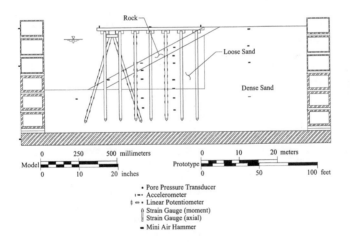

(b)   Wharf supported on a rock sliver fill (Centrifuge model JCB01).

Figure 1: Pile supported wharf configurations tested as a portion of this investigation.

the generation of excess pore pressures leading to liquefaction, the response of structural elements, and soil-structure interaction.

From the author's perspective a key limitation in the application of these methods in practice is the lack of rigorous validation of the procedures using case histories from ports. This is due largely to the limited number of well-documented case histories that exist for pile supported wharves and piers. The paucity of instrumentation for strong ground motion, slope deformation, and excess pore pressures at port facilities has significantly hindered research in this area. As a result port facilities are commonly analyzed using numerical tools for complex soil-foundation-structure interaction that may have been calibrated for dissimilar applications (e.g., bridge foundations, earth retention systems, embankment dams or slopes), which tends to increase the uncertainty associated with the computed results. This paper presents the results of an applied research effort conducted to compare computed wharf performance using the above analysis methods and the observed performance of field case histories and physical models. The physical models consisted of a suite of well-instrumented pile-supported wharf centrifuge models tested to supplement the limited field case history database (McCullough 2003, McCullough and Dickenson 2005). In addition, key issues regarding the seismic performance-based analysis of pile-supported wharves are presented.

METHODS OF ANALYSIS

The procedures most commonly employed in practice for estimating the permanent deformations of slopes, the rigid-body sliding block method and more sophisticated numerical continuum models, differ substantially in the effort required to obtain displacement estimates using suites of ground motion time histories considered appropriate for the regional seismic hazard. Project budgets often do not allow for in-depth numerical analysis therefore practitioners are confronted with balancing computational efforts with anticipated benefits. The sliding block method provides economy of effort with approximate estimates of rigid body displacement, while the more resource intensive numerical models are capable of providing approximate estimates of the patterns of deformations necessary for assessing pile performance. It is considered very worthwhile to demonstrate when the methods are suitable and appropriate. This study focuses on the application of two approaches for estimating the seismic performance of marginally stable slopes with embedded piles; 1) the rigid-body, sliding block method, and 2) a practice-oriented 2D non-linear effective stress model utilizing the commercially available program FLAC. These two procedures are used to compare computed slope deformations with those displacements measured in centrifuge models. Geotechnical parameters for both models were defined on the basis of measured data from extensive arrays of instrumentation in the centrifuge and on correlations with well defined soil parameters. Cursory aspects of the procedures employed in this investigation are described below (refer to McCullough 2003 for specific details of the modeling).

Rigid-Body, Sliding Block Method

Slope deformations have been estimated using the Newmark sliding block procedure based on simple limit equilibrium stability analysis. The application of the procedure has been well presented in the literature (e.g., Kramer 1996, Jibson 1993, Hynes-Griffin and

Franklin 1984, Makdisi and Seed 1978). In this study the actual pore pressures measured in the centrifuge model were used to obtain representative soil strengths in the limit equilibrium model and the effects of the piles were accounted for in a straightforward manner. The acceleration time histories used in the sliding block analyses were also recorded in the centrifuge model. Several pertinent strengths and limitations of this approach are outlined, with supplementary information on the selection of geotechnical parameters. The primary limitations of the sliding block method as applied in practice for the forecasting of seismic performance are: 1) the soil, particularly in the liquefiable zones, is not a rigid material, as assumed by the method; 2) the method is dependent on the acceleration time histories used as input, often with large variations due only to the characteristics of the motions (even when scaled to equivalent peak accelerations); 3) the pattern of slope deformations is not predicted; 4) a single mode of failure has to be assumed to calculate the critical acceleration; and 5) the forces/displacements applied from the soil to the pile elements are not readily computed. These limitations are balanced by practical advantages such as: 1) characteristics of the earthquake motions are captured (e.g. duration, frequency content, etc.) as representative design acceleration time histories are used; 2) the interaction of the piles and the soil can somewhat be accounted for, if only in a simplified manner, either as a resisting shear force or a pseudo-cohesion in the soils; 3) soil degradation can be accounted for as an equivalent shear strength for liquefied soils and post-cyclic reduction in strength for soft clays; and 4) deformations are easily predicted, which are often related to target performance criteria.

In the analyses presented herein the limit-equilibrium slope stability program UTEXAS3 (Wright 1992) was utilized for estimating the "post-cyclic", or "post-earthquake", static factor of safety, from which the yield coefficient was computed. The cyclic shear strength of the cohesive soil was assumed to be the same as the undrained static shear strength (i.e. the reduction in strength due to degradation was assumed to negate the increase in strength to the loading rate). This appears reasonable for the reconstituted San Francisco Bay Mud used in the models. The shear strength of the cohesionless soils was modified to account for any excess pore pressure that was measured in the centrifuge models. The shear resistance of fully liquefied sands was modeled using a residual undrained strength procedure (Stark and Mesri 1992, Olson and Stark 2002). The piles were included in the analyses as reinforcement elements. The strength of the piles was estimated using the method proposed by Broms (1964). The ultimate resistance of the soil as determined from Broms (1964) was applied as a reinforcement element shear resistance. The maximum shear resistance was determined to be the minimum value as determined assuming short pile or long pile behavior, as the short pile behavior is based on the soil capacity and the long pile method is based on the pile capacity, the minimum of which was assumed to be the actual capacity. The procedure was as follows:

1) The critical failure surface was estimated using an acceleration coefficient for which the factor of safety against slope stability was unity, without the pile elements.
2) The critical failure surface was then used to estimate the ultimate resistance from the piles using the assumptions noted above.
3) The ultimate resistance was then included in the same analysis as step 1; however, the resistance provided by the pile elements was now included, and a new critical failure surface was determined using an acceleration for which the factor of safety against slope stability was unity.

4) Steps 2 and 3 were then repeated until there was negligible difference in failure surface locations.

The resulting acceleration value determined in step 3 was then used as the yield acceleration, and the portion of the acceleration time histories exceeding the yield acceleration were double integrated to provide an estimate of the permanent slope deformation.

Numerical Analyses using a 2D Nonlinear, Effective Stress Model

The numerical modeling component of the investigation utilized the commercially available computer program FLAC (Itasca 2000). FLAC (Fast Lagrangian Analysis of Continua) is a two-dimensional finite-difference computer program developed for the modeling of geo-mechanical problems, and is well suited for the analysis of large strain problems such as liquefaction-induced ground failures. The validation and use of FLAC for the seismic modeling of port structures has been documented by Roth et al. (1992), McCullough et al. (2001), McCullough (2003), and McCullough and Dickenson (1998). The constitutive model consists of simple, and relatively easily calibrated, components including Mohr-Coulomb model in which the soil strength was defined by the cohesion and angle of internal friction, the volumetric strain behavior was defined by a dilation angle, and the elastic behavior was defined by the shear and bulk modulus of the soil. In addition, the constitutive model included a plastic flow rule to model plastic soil behavior. The implementation of this procedure in a constitutive model was originally adapted for FLAC by Roth et al. (1991) for the seismic analysis of Pleasant Valley Dam. A complete description of the constitutive model can be found in Itasca (2000), and specific aspects of the modeling as applied to these wharf studies in McCullough (2003).

The basic Mohr-Coulomb model was also modified to account for the cyclic generation of excess pore pressures using the cyclic stress approach, in which a damage parameter was related to the excess pore pressure. Dawson et al. (2001) provide an in-depth overview of a constitutive model that is essentially identical to the one described above. The numerical model included the effect of pore pressure generation, but pore pressure dissipation was not modeled. Liquefied soil strengths were modeled, using the recommendations of Stark and Mesri (1992). Ground surface settlements due to the consolidation of liquefied soils were modeled using a post-processing routine implementing the volumetric strain model of Ishihara and Yoshimine (1992).

The interaction of the piles and soil was modeled with SSI springs. The SSI springs had a strength (represented by a *cohesion* and a *friction angle*) and a stiffness. The properties of the SSI springs in the normal direction were determined using a model of the piles in plan view, at several different depths (stress states). The cross-section of the piles was pushed into the soil and the soil-structure response was used to estimate the spring stiffness and strength, following a procedure outlined by Itasca (2000). The SSI springs in the shear direction were estimated assuming that the shear stiffness was approximated by the shear modulus of the soil and the shear strength was estimated using estimates of skin friction commonly conducted for vertical pile capacity analysis.

The seismic boundary conditions for the field case history outlined in this paper (Port of Oakland) consisted of a free-field boundary condition, as implemented in the numerical model. The seismic damping in the models utilized the Raleigh damping option. For all of the models the percentage damping was chosen as 5 percent, based on typical soil behavior. The damped center frequency was determined by observing the undamped model behavior. The boundary conditions for centrifuge models were established to match the behavior of the laminar box used in the testing.

RESULTS OF THE SLIDING BLOCK VALIDATION STUDY

The rigid, sliding block procedure previously above was used in an attempt to replicate the slope displacement measured in the centrifuge models of pile supported wharves. Each centrifuge model was tested with a suite of earthquake motions, but only the first large earthquake motion (peak horizontal acceleration greater than 0.1 g) was used for the validation study, as each subsequent test introduced additional uncertainty in geometry and soil properties. Two of the five centrifuge models are shown in Figure 1. The results of the validation study are presented in Figure 2.

The direct comparison of the computed rigid-body displacement, wherein only one uniform displacement is obtained, with the measured slope movement in a physical model that is free to deform throughout the soil mass requires that the deformation at one location in the physical model be specified. The centrifuge displacements (Figure 2) were measured at the ground surface immediately adjacent to the wharf deck. In stability analyses of several centrifuge models (e.g., JCB01 without piles, SMS01 with and without piles) the post-earthquake factor of safety was less than 1.0 resulting in infinite displacements by sliding block procedures. These values are illustrated with arrows on the top of the figure. The open symbols represent analyses that did not include the effect of the piles and the solid symbols represent analysis that included the effect of the piles. The analyses utilized the mean plus one standard deviation excess pore pressure ratio for each soil layer that was recorded in the centrifuge models. The vertical error bars represent the range of predictions using the various recorded acceleration time histories. The horizontal error bars represent the range of the measured deformations between the wharf deck and adjacent backland soil.

The Newmark sliding block analyses utilized all of the earthquake acceleration time histories that were recorded within the approximate failure mass. This included between 7 and 14 time histories for each test. The range of predicted values, due to the use of the different acceleration time histories, is illustrated by the vertical error bars in Figure 2. The validation study shows that a reasonable prediction of deformations can be obtained using the standard sliding block method provided that the stability analyses account for excess pore pressure generation in the soils, degradation of soil strength, and the stabilizing effect of the piles. In general, the slope displacements were over-predicted when the effect of the piles was neglected, and the deformations were under-predicted when the effect of the piles was incorporated in the analyses. In addition, the best predictions were obtained when the mean plus one standard deviation pore pressure measured in the soil layers was utilized.

A primary uncertainty in displacement estimates made using the rigid, sliding block method in practice lies in accurately estimating the pore pressure generation throughout the soil mass, and computing representative acceleration time histories at the base of the failur plane and within the slide mass. These are usually not known *a priori*, as they were in thes validation studies using measured data from the centrifuge models. It is apparent from thi work that a significant range of computed displacements can occur even when these value are known. The goal is to therefore to bracket the likely range of displacements. Care shoul be taken when interpreting the results of rigid, sliding block analyses when used to estimat permanent lateral deformations.

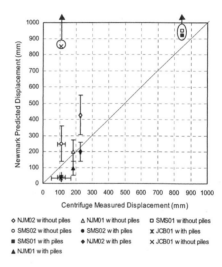

Figure 2:   Results of the rigid, sliding block validation study.

RESULTS OF THE NUMERICAL MODEL VALIDATION STUDY

Numerical continuum models provide numerous enhancements over rigid body, limi equilibrium procedures for the simulation of seismic wharf response; however, th complexities associated with the modeling require careful validation prior to project specifi applications. Some of the more pertinent modeling issues include; soil-pile interaction i liquefiable soil, pile behavior in sloping rock fill where particle sizes approach the diamete of the piles, piles with different unsupported lengths, pile performance at soil layer interfaces pore pressure generation and dissipation, and large-strain soil behavior. Experience ha demonstrated the significant influence that these modeling issues can have on computed soi deformations and pile performance. This investigation included an extensive validation an calibration exercise using the results of field case history data and centrifuge results on singl piles prior to applications involving the complete wharf models. A summary of the primar results of these studies is provided in the following sections of the paper.

Lateral Pile Behavior in Sloping Rock Fill

An initial validation was conducted to estimate the lateral response of piles in sloping rock fill. The measured performance of two centrifuge models (Figure 3) was used to assess the capability of the model to simulate lateral pile response to static loading. Two issues were evaluated in this stage of the project; particle size effects, and the influence of sloping ground on the lateral soil resistance adjacent to the pile. It was anticipated that conventional p-y relationships based on sands would not be applicable without modification to account for scale effects. The size of the rock particles relative to the pile diameter is such that a discrete element model would be preferable over the standard continuum–based models used for later pile response. The continuum assumption for soils is generally valid when the particle dimensions are on the order of 30 to 40 times less than the dimension of interest (for this case the pile diameter). For the rock fill, individual rock particles were approximately 3 times smaller than the pile diameter, clearly indicating that the rock fill was not a continuum in relation to the piles. In order to account for the particle interaction effects in a simple manner consistent with routine p-y type analyses, the best comparison between the numerical analyses and the measured lateral pile response was obtained when a *"pseudo-cohesion"* was applied for the rock fill. The pseudo-cohesion accounted for the individual rock particle/pile interaction in a simplified manner that was not accurately captured using the continuum model directly. The best fit for these models was found using the pseudo-cohesion of 15 kPa. In addition, it was found that the downslope normal spring stiffness was approximately a factor 10 less than the upslope stiffness. The predicted and measured responses for one load cycle are shown in (Figure 4), comparing the results using the noted modifications.

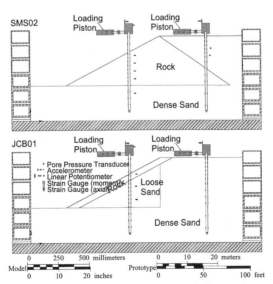

Figure 3:   Centrifuge Model Geometries for Lateral Pile Load Tests.

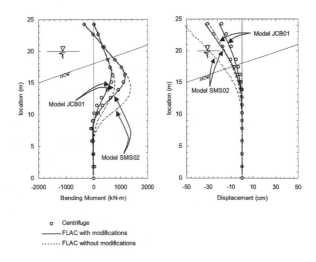

Figure 4:   Comparison between the predicted moments and displacements for the pile in the sloping rock fill.

Field Case History: Port of Oakland

The early results from the static centrifuge tests were used to calibrate the numerical model in preparation for simulations of a well documented field case study. The seismic performance of the pile supported wharves at the Seventh Street Terminal (Berths 35 - 38), Port of Oakland during the 1989 Loma Prieta Earthquake was simulated using the numerical model. Field observations indicated permanent ground surface lateral displacement of the rock dike on the order of 15 to 30 cm, with approximately 13 to 30 cm of settlement (Egan et al. 1992, Serventi 2003, Singh et al. 2001). In addition, the majority of the batter piles and approximately 20 percent of the vertical piles failed at the pile/deck connection (Singh et al. 2001). It was also noted by Singh et al. (2001) and Oeynuga (2001) that many of the vertical piles probably failed at the approximate interface between the Bay Mud/hydraulic fill and the dense sand, based on the results of pile integrity testing. These failures were likely due to pinning of the piles in the dense sands (i.e., stiffness contrast) while lateral forces due to permanent ground deformations pushed on the upper portions of the piles in the rock fill. Berth 38 was modeled with the design geometry shown in Figure 5 and numerical model grid, soil layers, and structural elements shown in Figure 6.

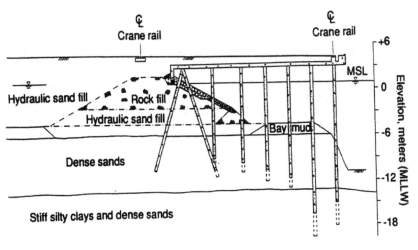

Figure 5:   Pile-supported wharf at the Port of Oakland Seventh Street Terminal, prior to the 1989 Loma Prieta Earthquake (Egan et al. 1992)

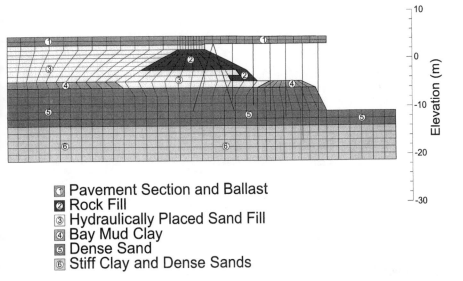

① Pavement Section and Ballast
② Rock Fill
③ Hydraulically Placed Sand Fill
④ Bay Mud Clay
⑤ Dense Sand
⑥ Stiff Clay and Dense Sands

Figure 6:   Geometry, grid and structural elements that were used in the numerical model of the Port of Oakland Seventh Street Terminal, Berth 38.

The nearest recorded acceleration time history was recorded at the ground surface approximately 1.5 kilometers from Berth 38 at the Port of Oakland Outer Harbor Wharf (Berth 24/25). The two horizontal components of the recorded motions were combined to produce a motion perpendicular to Berth 38. The combined motion was deconvolved to the base of the numerical model (El. –21 m) using SHAKE91 (Idriss and Sun 1992). The results of the numerical analysis are illustrated graphically in Figure 7. The computed horizontal displacement of the rock dike at the ground surface of 27 cm and a vertical settlement of 22 cm, agrees quite well with the observed values of 15 to 30 cm and 13 to 30 cm, respectively. In addition, plastic hinge development was indicated at the top of the all the piles (at the location of the first structural node below the wharf deck). In addition, plastic hinge development at depth was predicted (Figure 7) at the Bay Mud/hydraulic fill and dense sand interfaces.

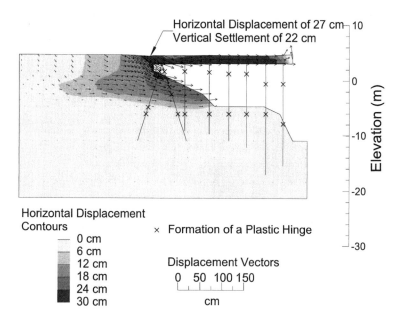

Figure 7:   Numerical simulation of the seismic performance of Berth 38 at the Port of Oakland during the 1989 Loma Prieta Earthquake.

Centrifuge Models

The numerical model was applied for all of the centrifuge tests examined as a part of the sliding block study. A complete description of the centrifuge models can be found in McCullough (2003) and McCullough and Dickenson (2005). Nevada sand was used and density measurements were taken during placement of the sand, and were related to the porosity, static strength, and cyclic strength, using the results of the laboratory tests for the VELACS study (Arulmoli et al. 1992). The rock strengths were estimated based on rock fill strengths published in the literature using the measured densities and porosities (Leps 1970, Marachi et al. 1972). The Bay Mud density and porosity were computed from water content tests of saturated samples of the clay, and the undrained strength values were measured using Torvane and Pocket Penetrometer tests in the clay, confirmed with the established undrained strength ratio for the clay.

The modulus values of the cohesionless soils were calculated based on in-situ (i.e., in-flight) shear-wave velocity tests. A miniature air hammer (Arulnathan et al. 2000) was used to generate shear waves within the centrifuge models, which were recorded by the vertical accelerometer arrays. Low strain shear modulus values were then back-calculated from the shear wave velocities. The undrained elastic modulus ratio (E_u/s_u) was estimated as 450 for the Bay Mud. Since the utilized constitutive model did not include cyclic modulus reduction, the modulus values of all the soils were reduced by 20 percent during the dynamic analyses to partially represent dynamic modulus reduction of the soils. Twenty percent represents the average reduction based on the average cyclic shear strain measured in the numerical model.

Two sizes of piles were used in the centrifuge models. The following properties are reported in prototype units. Models NJM01, SMS02 and JCB01 used aluminum tubing that was 64 cm diameter and 3.6 cm wall thickness, while NJM02 and SMS01 used aluminum tubing that was 54 cm diameter and 3.6 cm wall thickness. The modulus of elasticity of the aluminum was 70 GPa. The plastic moment of the piles was 7.5 MN-m and 48.7 MN-m for the 54 cm and 64 cm diameter piles, respectively. The SSI interaction springs were calculated for each model, at approximately 1 to 2 m intervals along the length of the piles.

The results of the physical and numerical modeling comparison study are summarized in Figure 8. The data provided in these plots is obtained from a series of five centrifuge models involving different rock fill configurations and pile orientations (McCullough 2003). The following general trends are evident:

1) The maximum and residual pile moments were generally not well predicted. The R^2 correlation coefficients for the maximum and residual moments were 0.22 and 0.16, respectively. Computed maximum moments tend to be greater than the measured values.

2) The pore pressures and accelerations were somewhat well predicted, with R^2 correlation coefficients of 0.58 and 0.60, respectively. The computed accelerations were generally larger than those measured in the centrifuge, and this is anticipated to be due to the lack of a strain-based modulus reduction scheme at small to moderate strain levels and the simplified method of damping that was used in the models.

3) The displacements were very well predicted, with an R^2 correlation coefficient of 0.89.

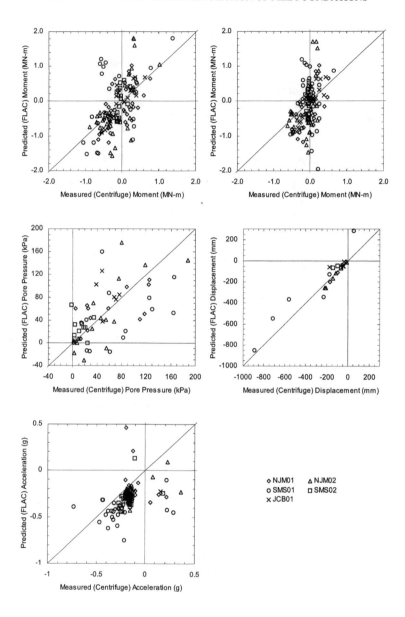

Figure 8:   Results of the FLAC validation study using the centrifuge model test data.

It should be noted that numerous sensitivity analyses were conducted to evaluate the effect of various modeling parameters (e.g. soil-pile interaction spring stiffness and strength, numerical model damping, changes in shear modulus as a function of pore pressure generation and cyclic strain, etc.). The results presented in Figure 8 represent analyses that best fit the measured centrifuge performance. The generally poor prediction of the computed pile moments is not well understood although several possible limitations of the soil-pile interaction as treated by the numerical model are provided. The numerical model accounted for the rock particle behavior, as well as the difference in upslope and downslope stiffness. It was anticipated that the maximum pile moments would be fairly well modeled provided that the ground and wharf deck accelerations were representative. This is due to the fact that the maximum moments are often observed to occur immediately beneath the wharf deck and related to the inertial loading associated with the deck. It was also anticipated that the computed residual moments in the piles would be accurately modeled if the magnitude and pattern of soil displacements were well predicted given the significance of permanent ground deformations on pile curvature. The displacements and accelerations were well reproduced; however, there is considerable uncertainty in the computed pile moments.

A comparison of the measured and computed residual moments in the piles for centrifuge model SMS01 is provided in Figure 9. It can be seen that the numerical model largely over-predicted the moments at the soil interfaces (especially for pile number 2). It is interesting to note the large bending moments that developed at depth within the piles were approximately the same magnitude as the moments at the top of the piles. These large moments would not be anticipated using standard depth of fixity methods for lateral pile response as the layer interfaces at which the moments occur can be beneath the computed fixity depth. The numerical model tended to over-predict the moments at depth thereby providing a conservative result. This is in contrast to analysis methods that model the soil as a set of springs fixed in space that are attached to the piles. These procedures do not account for deep seated global soil deformations or strain profiles that develop near soil interfaces.

SUMMARY AND CONCLUSIONS

The configuration of common pile supported wharves combined with geotechnical conditions that often include weak soils contribute to a challenging dynamic soil-foundation-structure interaction problem. Current geotechnical tools are capable of reliably simulating the seismic performance of these structures; however, they should be rigorously validated prior to use in practical applications. This investigation involving integrated modeling of field case studies, physical modeling in the geotechnical centrifuge, and validation of common procedures for estimating wharf and embankment performance was undertaken to address uncertainties inherent in the methods. Pile performance has been observed to be directly related to the magnitude of ground failures at waterfront sites. Structural failures of piles and pile-wharf deck connections are often in response to ground failures (slope instability, liquefaction, etc.). The performance-based design of pile-supported wharves requires a clear understanding of the seismic geotechnical performance. The lack of well instrumented field case histories has resulted in a situation where analysis methods for estimating the geotechnical performance have undergone limited validation. The results of this validation study indicate that seismically-induced deformations can be well predicted provided that pore pressure generation and the effects of the piles (sometimes referred to as "pile pinning") is adequately addressed, if even in a simplified manner. Of primary importance in practice, it is apparent that the prediction of maximum and residual moments in piles is associated with considerable uncertainty. This uncertainty should be acknowledged in efforts to satisfy performance-based design criteria. This work highlights the necessity of coordinated analysis by both geotechnical and structural engineers in order to assess the performance of piles.

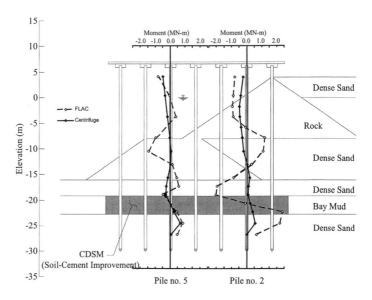

Figure 9:   Comparison between the Centrifuge measured and FLAC predicted residual moments in piles number 2 and 5 for centrifuge model SMS01.

The results of these analyses demonstrate the necessity of model validation and calibration against field case history data and/or physical modeling. A lack of validation can lead to a poor understanding of the uncertainty involved in seismic analyses, and an over-confidence in the results. Several key issues pertaining to the modeling of wharf performance have been gleaned from this investigation:

1) Representative estimates of permanent, seismically-induced deformations of pile-supported wharves can be made using the rigid, sliding block method of analysis provided that the following issues are addressed; the stabilizing effect of the piles, pore pressure generation throughout the soil masses, and soil strength degradation. It should be acknowledged that computed deformations will still vary in response to the characteristics of the acceleration time histories that are used, and method of accounting for the possible reduction in yield acceleration during cyclic loading.

2) The location of the critical circle identified in limit equilibrium slope stability analyses will change with incorporation of piles or soil improvement. If the analysis proceeds in steps that include the soil only, then the soil with piles, and possibly soil-piles-soil improvement the influence of these factors should not be assessed by comparing the factors of safety obtain for the same critical circle. The stability analyses must be performed so that the new critical circle is obtained for each case. This will affect the yield acceleration and the corresponding displacement estimates.

3) Cyclic lateral behavior of piles in sloping rock fill can be fairly well modeled using a continuum model if the difference between upslope and down slope SSI spring stiffness is accounted for. In addition, it appears that one simplified approach for handing the influence of scale effects in p-y relationships for lateral loading of piles in rock fill is the use of a *pseudo-cohesion* for the rock fill to account for the individual rock particle interaction (i.e., interlocking, rotation).

4) The numerical model used in this study was found to provide reasonable estimates of seismically-induced permanent deformations, ground and structural accelerations, and excess pore pressure generation. This was true despite the use of the Mohr-Coulomb constitutive model in conjunction with a simplified stress-based pore pressure generation model to represent dynamic pore pressure generation. The advantage of using these simple and more approximate methods over more sophisticated constitutive models is that standard geotechnical parameters are used and the results can be quickly verified. Limitations were however observed in the modeling as performed herein. Both the maximum cyclic and residual moments in piles were largely over estimated, even though the displacements, pore pressures and accelerations were fairly well modeled. These uncertainties should be acknowledged in practice. It is anticipated that improvements in the moment estimates could be made with modifications to the numerical model. Specific topics of future work include; improved modeling of near-pile soil behavior, pore pressure dissipation throughout the model, the incorporation of fully nonlinear soil-pile springs (p-y response) for liquefied soil, and an improved strain hardening model for soils subjected to large strain. These enhancements notwithstanding it must be acknowledged that uncertainties will continue to exist when using 2D models to simulate highly 3D geomechnical problems such as piles in slopes.

5) This work complements observations made in the field that large pile moments can develop at depth due to global soil displacement or large stiffness contrasts at soil interfaces. Models capable of simulating the patterns of soil deformation at depth are required to reliably estimate these moments. The application of displacement profiles from simplified displacement methods into structural lumped-mass models for pile performance are uncoupled and fail to capture important aspects of dynamic SFSI.

6) The use of scaling factors for the summation for forces in piles due to the kinematic and inertial components of loading is not recommended at this time. This approach has been applied in practice for bridge foundations and waterfront structures; however, the procedure requires a judicious selection of weighting ratios. The timing of peak loads induced by the kinematic and inertial components of loading are not concurrent therefore the direct summation of weighted loads is not a straightforward process. The phasing of these loads must be accounted for in any procedure that is based on a simplified weighting scheme.

ACKNOWLEDGEMENTS

This research was supported by Grant No. CMS-9702744 from the National Science Foundation, and Grant No. SA2394JB from the Pacific Earthquake Engineering Research Center. Valuable industry support of the project was provided by the Port of Oakland, Port of Long Beach, URS Corp, and Hayward Baker, Inc. The support from these agencies and firms is greatly appreciated. The authors would like to gratefully acknowledge Scott Schlecther (Geotechnical Resources, Inc.) and Jonathan Boland (Raney Geotechnical, Inc.) for their tremendous contributions in the construction and testing the centrifuge models. Also Bruce Kutter, Dan Wilson, and the staff of the Geotechnical Centrifuge Facility, Center for Geotechnical Modeling at UC Davis for their substantial assistance and valuable insights during the construction and model testing. Finally, to Manfred Deitrich and Andy Brickman at Oregon State University for their efforts in the fabrication of models and instrumentation.

REFERENCES

Arulmoli, K., Muraleetharan, K.K., Hossain, M.M., and Fruth, L.S. 1992. *VELACS: Verification of Liquefaction Analyses by Centrifuge Studies Laboratory Testing Program Soil Data Report.* Prepared for NSF by The Earth Technology Corp. Proj. No. 90-0562.

Arulnathan, R., Boulanger, R.W., Kutter, B.L., Sluis, B. 2000. "A New Tool for V_s Measurements in Model Tests." *ASTM Geotechnical Testing Jrnl.* Vol. 23, no. 4, 444-452.

Broms, B.B. 1964. "Lateral Resistance of Piles in Cohesionless Soils." *Journal of the Soil Mechanics and Foundations Division.* American Society of Civil Engineers. Vol 90, No. SM3, May, 1964. pp 123-157

Dawson, E.M., Roth, W.H., Nesarajah, S., and Davis, C.A. 2001. "A Practice Oriented Pore-Pressure Generation Model." Proceedings of the 2nd *FLAC Symposium*, Lyon, France. A.A. Balkema Publishing, Rotterdam, Netherlands.

Egan, J.A., Hayden, R.F., Scheibel, L.L., Otus, M., and Serventi, G.M. 1992. "Seismic Repair at Seventh Street Marine Terminal," Proc. of *Grouting, Soil Improvement, and Geosynthetics*, ASCE Geotechnical Speciality Publication No. 30, Vol. 2.

Hynes-Griffin, M.E., and A.G. Franklin. 1984. "Rationalizing the Seismic Coefficient Method," U.S. Army Corps of Engineers Waterways Experiment Station. *Miscellaneous Paper GL-84-13.* Vicksburg, MS. 21 p.

Idriss, I.M., and Sun, J.I. 1992. "User's Manual for SHAKE91 – A Computer Program for Conducting Equivalent Linear Seismic Response Analyses of Horizontally Layered Soil Deposits," Center for Geotechnical Modeling, Univ. of California, Davis.

Ishihara, K., and Yoshimine, M. 1992. "Evaluation of Settlements in Sand Deposits Following Liquefaction During Earthquakes." *Soils and Foundations.* Japanese Society of Soil Mechanics and Foundation Engineering. Vol. 32, No. 1. Pp. 173-188.

Itasca. 2000. *FLAC version 4.0 User's Manual.* Itasca Consulting Group, Minneapolis, MN.

Jibson, R.W. 1993. "Predicting Earthquake-Induced Landslide Displacements Using Newmark's Sliding Block Analysis." *Transportation Research Record, No. 1411- Earthquake-Induced Ground Failure Hazards.* TRB, NRC. Washington, D.C. pp. 9-17.

Kramer, S.L. 1996. *Geotechnical Earthquake Engineering.* Prentice Hall, Upper Saddle River, NJ. 653 pp.

Leps, T.M. 1970. "Review of Shearing Strength of Rockfill." *Journal of the Soil Mechanics and Foundations Division.* ASCE. Vol. 96, No. SM4. pp. 1159-1170.

Makdisi, F.I. and Seed, H.B. 1978. "Simplified Procedure for Estimating Dam and Embankment Earthquake-Induced Deformations." *Journal of the Geotechnical Engineering Division.* ASCE. Vol. 104, No. GT7. pp. 849-867.

Marachi, N.D., Chan, C.K., and Seed, H.B. 1972. "Evaluation of Properties of Rockfill Materials." *Journal of the Soil Mechanics and Foundations Division.* ASCE. Vol. 98, No. SM1. pp. 95-114.

McCullough, N.J. 2003. "The Seismic Geotechnical Modeling, Performance, and Analysis of Pile Supported Wharves," PhD Dissertation, Dept. of Civil, Construction and Environmental Engineering, Oregon State University, 204 p.

McCullough, N.J. and Dickenson, S.E. 1998. "Estimation of Seismically induced Lateral Deformations for Anchored Sheetpile Bulkheads." Proceedings of the *Geotechnical Earthquake Engineering and Soil Dynamics III* conference, Seattle WA. P. Dakoulas, M. Yegian, and R.D. Holts (eds.). Geotechnical Special Publ. No. 75. ASCE, pp 1095-1106.

McCullough, N.J. and Dickenson, S.E. 2005. "The Dynamic Centrifuge Modeling of Pile-Supported Wharves." In review, *ASTM Geotechnical Testing Journal.*

McCullough, N.J., Dickenson, S.E., and Pizzimenti, P.B. 2001. "The Seismic Modeling of Sheet Pile Bulkheads for Waterfront Applications." Proceedings of the *2nd FLAC Symposium*, Lyon, France. A.A. Balkema Publishing, Rotterdam, Netherlands.

Olson, S.M. and Stark, T.D. 2002. "Liquefied strength ratio from liquefaction flow failure case histories." *Canadian Geotechnical Journal.* Vol. 39, No. 3, pp. 629-647.

Oyenuga, D., Abe, S., Sedarat, H., Krimotat, A., Slalah-Mars, S., and Ogunfunmi, K. 2001. "Analysis of Existing Piles with Missing Data in Seismic Retrofit Design at the Port of Oakland," *Proc. of the ASCE Ports 2001 Conference*, Norfolk, VA, paper on CDROM.

PIANC (International Navigation Association). 2001. *Seismic Design Guidelines for Port Structures.* International Navigation Association Working Group No. 34. A.A. Balkema.

Roth, W.H., Bureau, G., Brodt, G. 1991. "Pleasant Valley Dam: An Approach to Quantifying the Effect of Foundation Liquefaction." *Proceedings of the 17th International Congress on Large Dams.* Vienna, Austria. pp. 1199-1223.

Roth, W.H., Fong, H., and de Rubertis, C. 1992. "Batter Piles and the Seismic Performance of Pile-Supported Wharves." *Proceedings of the '92 Ports Conference.* American Society of Civil Engineers. Seattle, WA, July 20-22, 1992. pp 336-349.

Serventi, J. 2003. "Port of Oakland: Lessons Learned from the Loma Prieta Earthquake." Presented at the seminar on the *Geotechnical Engineering for Waterfront Structures.* Sponsored by the Seattle Section Geotechnical Group of American Society of Civil Engineers, Seattle, WA. March 15, 2003.

Singh, J.P., Tabatabie, M., and French, J.B. 2001. "Geotechnical and Ground Motion Issues in Seismic Vulnerability Assessment of Existing Wharf Structures," *Proc. of the ASCE Ports 2001 Conference*, Norfolk, VA, paper on CDROM.

Stark, T.D. and Mesri, G. 1992. "Undrained Shear Strength of Sands for Stability Analysis." *Journal of Geotechnical Engineering.* ASCE. Vol. 118, No. 11. pp. 1727-1747.

Werner, S.D. (Editor). 1998. *Seismic Guidelines for Ports.* ASCE Technical Council on Lifeline Earthquake Engineering. Monograph No. 12. March 1998. ASCE, Reston, VA.

Wright, S.G., 1992. *UTEXAS3, A Computer Program for Slope Stability Calculations.* Austin, Texas. May, 1990, revised July 1991 and 1992.

NONLINEAR ANALYSES FOR DESIGN OF PILES
IN LIQUEFYING SOILS AT PORT FACILITIES

Yoshiharu Moriwaki, Member, ASCE[1], Phalkun Tan, Member, ASCE[2],
Yoojoong Choi, Associate Member, ASCE[3]

ABSTRACT

An overview of nonlinear approach currently used by one group of practitioners in designing foundations for port facilities involving piles through potentially liquefiable soils is presented. The computer program FLAC is used as the overall model in this nonlinear approach, but other programs can be used as well. The approach consists of extensive parametric analyses using relatively simple models anchored in the knowledge of fundamentals and experience evaluating case histories. Given a postulated input ground motion, some parameters in the models are specified based on one-dimensional analysis using SHAKE and FLAC. Then, parametric gravity turn on analysis and parametric dynamic analysis are performed on a two-dimensional section without piles using pre-liquefied soil strength. Next, the section with piles is analyzed first using pre-liquefied soil shear strength. This process and the results from these analyses are used to develop required geotechnical design parameter values. Finally, the designed section is analyzed modeling an excess pore water pressure generation (liquefaction process) and design parameter values are adjusted as necessary.

INTRODUCTION

We now have commercially affordable computer programs capable of nonlinear analysis and increasingly fast inexpensive personal computers. When these are

---

[1] Principal, GeoPentech, Santa Ana, CA.
[2] Associate, GeoPentech, Santa Ana, CA.
[3] Senior Staff Engineer, GeoPentech, Santa Ana, CA.

combined with increasing demands for performance based seismic design by structural and other engineers, we see increasing usefulness of nonlinear analysis in various geotechnical design processes including one for designing piles in liquefying soils at port facilities., Not too many years ago, almost all geotechnical design processes usually and often necessarily used some forms of simplified methods, perhaps with confirmatory nonlinear analysis in some cases. Although use of simplified analyses is still appropriate or even preferred for many situations, a design approach based on nonlinear analysis should be seriously considered for some problems.

The objective of this paper is to provide a description of nonlinear design approach we have been using (e.g., Johnson et al., 1998) by following an example case to illustrate parts of the approach. The example provided is a hypothetical case and by no means complete. The intent is to present a sense of what such a design process involves and raise an issue or two for consideration.

OVERVIEW OF DESIGN PROCESS

The overall nonlinear model is based on the computer program FLAC (Itasca, 2001) herein; however, any other commercially available well-documented nonlinear computer program should be adequate. Given the overall FLAC model ("FLAC"), the design process focuses on parameters having relatively large uncertainty using simple models. Such models for port sites include ground motions, stress-strain relationships for soils, properties of pile-soil connection, and discretization of section leading to kinematic considerations. Although there are many important issues associated with ground motions in nonlinear analysis, they are not addressed herein.

In general, uncertainty in these models and their parameters associated with "FLAC" is considered less of an issue when a certain approach is adopted: simple but adequate models that are reasonably consistent with the current state of understanding are used to perform adequate parametric analysis to develop understanding of the effects of these parameters on the seismic performance of the site under consideration. Experience and confidence are gained when such an analysis approach is used to evaluate a number of case histories. Although case history evaluations are vital in anchoring the process, they are not addressed herein other than to note that the approach described herein has been used to evaluate a number of case histories associated from the 1971 San Fernando earthquake (Moriwaki et al., 1998) to the 1995 Kobe earthquake. In applying the design process to a soil-pile-structure system representing a port facility, it is useful to keep in mind the uncertainty involved in the overall as well as the component models and the complexity of the overall system being evaluated. It is also vital that professionals performing nonlinear analysis have adequate understanding of the computer program and formulations used in various models used in the program. Such understanding often provides safeguard against errors in modeling and numerical problems that are often rather difficult to discover but affect the results in subtle manner.

Simple but Adequate Models

The stress-strain behavior of soils is modeled using the Mohr-Coulomb model available in FLAC (Itasca, 2001). The Mohr-Coulomb model in FLAC has been used in seismic deformation analysis of earth structures by many people (e.g., Roth et al., 1993; Moriwaki et al., 1998). The modulus in this model is specified based on one-dimensional response analysis discussed later, and the shear strength in this model is specified based on usually available geotechnical parameters such as effective friction angle and cohesion intercept or undrained strength depending on the situation.

For soils that can potentially liquefy, the values of modulus and shear strength are degraded as functions of excess pore water pressures generated during earthquake shaking. The seismically induced excess pore water pressures are estimated using a stress-based approach (Roth et al., 1993; Moriwaki et al., 1998) similar to those used by others involving use of familiar number of cycles versus cyclic stress plots (e.g., De Alba et al., 1975) together with the confining pressure dependency of such relationships. The values of modulus and shear strength reach certain prescribed values when excess pore water pressure reaches "liquefaction" level. Most analyses in the nonlinear design approach described herein use these liquefied values of modulus and shear strength from the beginning; thus, these analyses use "pre-liquefied" properties of soils. The use of pre-liquefied shear strength in analysis in significant part of parametric analysis is often considered reasonable at least in highly seismic areas such as many parts of California.

The pile elements in FLAC (basically scaled beam elements) are connected to soil elements by p-y curves, providing lateral transfer of forces between piles and surrounding soils. Unless it is for specific purposes, the "t-z" curves (axial direction) are usually ignored. A bilinear idealization of p-y curves initially based on API guideline is used. Both the "stiffness" part and the "strength" part of resulting p-y curves are degraded as functions of excess pore water pressures generated by the earthquake shaking to prescribed values (Wilson et al., 2000; Abdoun et al, 2003), which depend on initial looseness of the surrounding soils. Here again, most analyses use "pre-liquefied" p-y curves.

An analysis section is discretized considering dynamic transmission implications of seismic energy and kinematic constraints. The former allows the transmission of seismic energy associated with all the appropriate frequencies; the latter ensures that element sizes allow deformation of materials without providing unnecessary constraints.

Design Process

Once the design earthquake ground motions are postulated and input ground motions are specified, the nonlinear design process can be summarized by the following six steps:

1. Free-Field one-dimensional response analysis

    By performing one-dimensional SHAKE (Schnabel et al., 1972, Idriss and Sun, 1992) analysis and one-dimensional FLAC analysis of free-field columns assuming no liquefaction, values of initial modulus in the soil model is specified, Rayleigh damping values are selected, and element heights are specified. In this step, the one-dimensional responses of FLAC are "forced" to be reasonably compatible with SHAKE results in the equivalent linear sense in the absence of liquefaction and permanent displacements.

2. Parametric two-dimensional analysis on section without piles

    Next using a two-dimensional section without piles and assigning to potentially liquefiable soils a reasonable range of post-liquefaction shear strength (residual shear strength), parametric gravity-turn-on analysis is performed. Thus, in the first part of this step, one begins to understand the general deformation patterns of the section without piles and to make adjustments to element sizes from kinematic point of view. However, most importantly, one can avoid situations too close to gravity-induced large deformation of the section under the post-liquefaction conditions. This part of the process starts considerations for subsurface improvements if necessary, but the soil improvement issues are not discussed herein.

    The second part of this step involves performing parametric dynamic deformation analysis on the section without piles using various input motions and the range of pre-liquefied shear strength. This part of the step initiates understanding of seismically induced deformation levels and patterns and possibly necessitates refinements in element sizes or a possible inclusion of interface elements for kinematic considerations. The amounts of seismically induced deformation of soil slopes are in general relatively faster to compute and easier to estimate than that of slopes consisting of soils and piles. Further, one feels more confident of the results on soil slopes without piles than those on more complicated soil slopes with piles.

3. Parametric two-dimensional analysis on section with piles

    A subset of analysis cases used under step 2 is used to analyze a section with piles again usually using pre-liquefied shear strength values. This step often involves discussions with structural engineers. The main purpose of this step is to evaluate the effects of piles and, if necessary, to develop information to specify the size of piles or reinforcements in the piles.

4. Development of design parameter values based on understanding the results of previous steps

    At least parts of this step occur throughout the previous steps. The results of steps 1 through 4 in terms of range of response and improved understanding of mechanisms of deformations are used to develop design information for geotechnical, structural, and possibly other engineers. Typical design parameters are the type, size, and length of piles; and design displacements, deformations, moments, shears, and rotations likely to be experienced by piles under the postulated seismic shaking conditions,

5. Two-dimensional analysis using "simple" models on section with piles

    Either as part of Step 4 or as the confirmatory analysis, the designed section with piles are used to perform dynamic FLAC analysis incorporating the liquefaction process (seismically induced generation of excess pore water pressures). Depending on the nature of the project and the results of Steps 1 through 4, this step may involve single or multiple analysis cases.

6. Iteration of parts or all of the previous steps as necessary

    Because this is a design process, often parts of the steps may need to be adjusted and repeated. In some cases, parts of the steps may even be eliminated because perhaps the results are obvious or not pertinent.

On the basis of our experience in applying the above design process on dozens of projects, the results of the above process are considered to provide adequately robust geotechnical designs, given the current state of understanding of soil behavior and soil-pile interactions under liquefying environments and the general trend towards performance based design in geotechnical engineering. It is also found that the above design process does not imply huge increase in evaluation costs when compared to use of more simplified analyses particularly given the nature of design information generated.

EXAMPLE CASE

To illustrate a sense of the design approach based on "FLAC", the section shown on Figure 1 is used as an example to provide a sense of what the above-described steps entail. It is a hypothetical port site with a wharf supported on vertical piles. The material 2 in Figure 1 corresponds to a loose sand layer considered to be potentially liquefiable. The basic material properties of various materials are also listed in Figure 1. The only additional material property information other than liquefaction-related properties is shear wave velocity.

One-dimensional Free-Field Response Analysis

Given input ground motions in terms of acceleration time histories, one-dimensional SHAKE (Schnabel et al., 1972; Idriss and Sun, 1992) and FLAC analyses are performed using free-field soil columns of the analysis section. The response spectral values at the ground surface are usually used to compare the results of the two sets of one-dimensional site response analyses as illustrated on Figure 2. This analysis is performed assuming that liquefaction of soils does not take place throughout the soil column. From this analysis, element height, Rayleigh damping, and initial modulus associated with soils are specified. These parameters are specified in such a way to not only provide adequate response spectral values at the ground surface but also to try to minimize the computation time.

Mat'l No.	Description	Unit Weight	Cohesion	Friction	Residual Shear Strength
		(pcf)	(psf)	(deg.)	(psf)
1	Dense Sand Layer	119	100	30	-
2	Loose Sand Layer	117	0	32	1000
3	Dense Sand Layer	128	0	40	-
4	Riprap	140	0	45	-

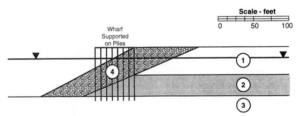

Figure 1: Example section and material properties

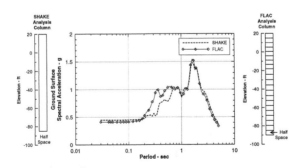

Figure 2: One-Dimensional site response analysis

Parametric Two-Dimensional Gravity-turn-on Analysis – Section without Piles

Figure 3 shows a section without piles that is discretized for two-dimensional FLAC analysis developed based partly on one-dimensional analysis. Computed displacements of the monitoring point about midway between the top and the bottom of the rip-rap slope shown on Figure 3 are used to provide a sense of both static and seismically induced displacements and deformations.

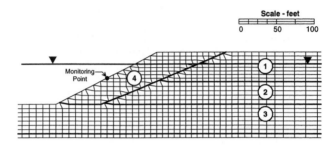

Figure 3: Discretized section without piles

Figure 4 presents the computed gravity-induced permanent displacements of the monitoring point shown on Figure 3 as a function of post-liquefaction shear strength assigned to soil in Layer 2. It is important to know that the slope involved is adequately far from gravity-induced slope instability, which occurs here at 200 psf.

As long as the mesh (Figure 3) needs to be generated for other analyses, more information can be gained from the FLAC based "slope stability" analysis than just from the slope stability analysis using limit equilibrium methods. Such information may include changes in the amounts and sometimes patterns of deformation as a function of, in this case, residual shear strength of soil in Layer 2.

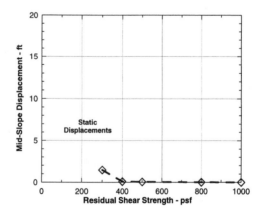

Figure 4: Results of paraetric gravity turn on analysis

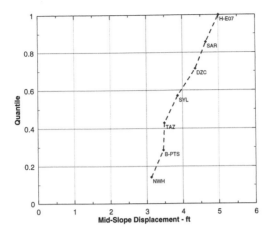

Figure 5: Quantile plot of computed displacements for various input time histories

Parametric Two-Dimensional Dynamic Analysis – Section without Piles

The analysis section without piles shown in Figure 3 is used in a series of dynamic analysis using seven different input time histories. The resulting computed displacements of the monitoring point are shown on Figure 5 as a quantile plot. The letters next to symbols on Figure 5 identify various input motions, all of which were adjusted in time domain to fit (Abrahamson, 1998) the postulated design response spectrum. Although issues on ground motions are not discussed herein, one should be aware of sensitivity in computed displacements caused by various input motions.

Figure 5 presents the computed seismically induced permanent displacements of the monitoring point shown on Figure 3 as a function of post-liquefaction shear strength assigned to soils in Layer 2. The same input motion is used in all the cases. The computed displacement rapidly increases as the residual shear strength is reduced from 1000 psf to 300 psf. A comparison of Figures 6 and 4 shows the effects of earthquake shaking on the computed displacement of the monitoring point.

Figure 7 shows a deformed mesh from one of the cases presented in Figure 6. Note that the deformed mesh reflects an exaggeration factor of 5. However, the deformed mesh indicates that the last row of elements at the bottom of Layer 2 shows significant deformation gradients, indicating that the element size may be imposing some kinematic constraints. To see the possible effects of this, interface elements representing the same material property with the soil in Layer 2 were used at the base of Layer 2, and the seismic deformation analysis was repeated for two cases. The resulting computed seismically induced displacement values of the monitoring point together with the results in Figure 6 are shown in Figure 8. It is clear from Figure 8 that the use of interface elements significantly increases the computed displacements.

The general deformation pattern shown on Figure 7 is often a result of assigning a constant value of shear strength of liquefied soil in a potentially liquefiable layer, a rather common current practice. These types of observations, which are not uncommon, provide additional uncertainty in the design process and require clarification from research findings.

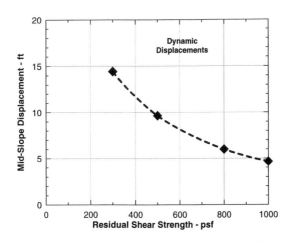

Figure 6: Effects of residual shear strength on computed seismic displacements

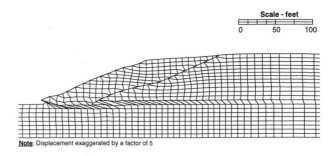

Figure 7: Deformed mesh from seismic deformation analysis

Parametric Two-Dimensional Analysis on Section with Piles

Figure 9 shows a deformed mesh at the end of shaking corresponding to a case of pre-liquefied shear strength of 1000 psf without use of interface elements. Figure 10 shows the computed deformation profiles of piles for the same case shown on Figure

9. Similar plots for rotations, moments, and shears can be developed for use by structural engineers if the results correspond to the desired design conditions.

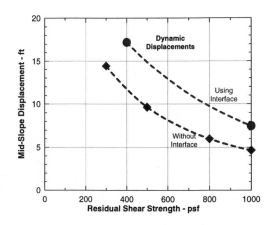

Figure 8: Effects of interface elements on computed seismic displacements

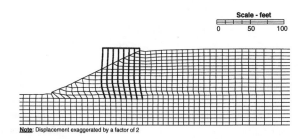

Figure 9: Seismically deformed mesh on section with piles

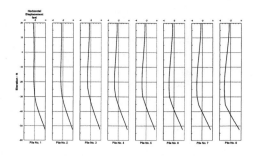

Figure 10: Computed seismic permanent deformation profiles of piles

Although in general piles and soils move by slightly different amounts during postulated earthquake shaking, they usually do not differ by very significant amounts. Without the piles the monitoring point moved about 5 feet (Figure 8), but with piles the monitoring point moved only about 2 feet. Figure 11 compares the pattern of deformed mesh for sections with and without piles. It is not clear in general whether such significant reductions in computed deformation amounts are reasonable or not. This is another example requiring clarification based on research findings.

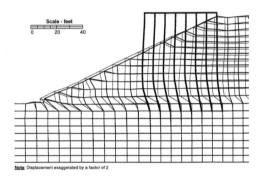

Figure 11: Comparison of computed seismic deformation patterns of sections with and without piles

SUMMARY AND CONCLUSIONS

An overview of nonlinear approach currently used by one group of practitioners in designing foundations for port facilities involving piles through potentially liquefiable soils is presented. The computer program FLAC is used as the overall model in this nonlinear approach, but other programs can be used as well. The approach consists of extensive parametric analyses using relatively simple constitutive models anchored in experience evaluating case histories. Given a postulated input ground motion, some parameters in the analysis are specified based on one-dimensional analysis using SHAKE and FLAC. Then, parametric gravity turn on analysis and parametric dynamic analysis are performed on a two-dimensional section without piles using pre-liquefied soil shear strength. Next, the section with piles is analyzed first using the pre-liquefied soil shear strength. The results from these analyses are used to develop a geotechnical design. Finally or sometimes as part of developing the final design, the designed section is analyzed modeling an excess pore water pressure generation (liquefaction process). The various performance results on piles usually from the last analysis are provided for use by structural engineers. Often at least parts of this process are repeated to refine the design.

It has been our experience that two-dimensional nonlinear analysis outlined herein is a useful design tool for piles in liquefying soils at port facilities. Use of "simple-but-no-simpler" models should be adequate in developing robust design when parametric analysis approach focusing on uncertain model parameters is used together with case history experience. Certain protocols such as one-dimensional response analysis of free-field should be followed for "QA/QC" of the results. In many design process, a number of questions are raised. These questions hopefully would be clarified by better dissemination of current or future research results.

REFERENCES

Abdoun, T., Dobry, R., O'Rourke, T. D., and Goh, S. H. (2003). "Pile response to lateral spreads: Centrifuge modeling." J. of Geotechnical and Geoenvironmental Engrg., ASCE, Vo. 129(10), 869-878.

Abrahamson, N.A. (1998). "RSPMATCH computer program." Personal communication.

De Alba, P., Chan, C. K., and Seed, H. B. (1975). "Determination of soil liquefaction characteristics by large-scale laboratory tests," Earthquake Engineering Research Center Report No. EERC 75-14 May 1975.

Idriss, I.M., and Sun, J.I. (1992). "SHAKE91: A computer program for conducting equivalent linear seismic response analyses of horizontally layered soil deposits," User's Guide, University of California, Davis, California, 13 pp.

Itasca (2001). Fast Lagrangian Analysis of Continue (FLAC) version 4.0 User's Guide and Manuals. August 2001.

Johnson, R. K., Riffenburgh, R., Hodali, R., Moriwaki, Y., and Tan, P. (1998). "Analysis and Design of a Container Terminal Wharf at the Port of Long Beach." Proc. of ASCE Ports 98, Long Beach, California, 436-444.

Moriwaki, Y., Tan, P., Ji, F. (1998). "Seismic Deformation Analysis of the Upper San Fernando Dam under the 1971 San Fernando Earthquake", Proc. of Geotechnical Earthquake Engineering and Soil Dynamics Conference, Geotechnical Special Publication No. 75, Seattle, Washington, 854-865.

Roth, W. H., et al. (1993). "Upper San Fernando Dam 1971 Revisited," 1993 Annual Conference Proceedings of the Associations of State Dam Safety Officials, Lexington, Kentucky: Assoc. State Dam Safety Officials, 49-60.

Schnabel, P.B., Lysmer, J., and Seed, H.B. (1972). "SHAKE, A Computer Program for Earthquake Response Analysis of Horizontally Layered Sites", EERC Report No. 72-12, University of California, Berkeley.

Wilson, D. W., Boulanger, R. W., and Kutter, B. L. (2000). "Observed Seismic Lateral Resistance of Liquefied Sand", J. of Geotechnical and Geoenvironmental Engrg. Vol. 126 (10), 898-906.

NUMERICAL ANALYSIS OF RATE-DEPENDENT REACTION OF PILE IN SATURATED OR LIQUEFIED SOIL

Ryosuke Uzuoka[1], Noriaki Sento[1] and Motoki Kazama[1]

ABSTRACT

The purpose of this study is to clarify the loading rate dependency of subgrade reaction of a single pile in saturated or liquefied soil. The parametric studies were performed by incorporating the soil-pore water coupled formulation and a newly proposed constitutive model for sand. The model can explicitly treat the degree of liquefaction by changing the lower limit of the mean effective stress. The liquefied soil at a certain depth around a pile is modeled with finite elements under plane stress condition. The subgrade reaction of a pile is calculated under various loading frequencies, permeability and degrees of liquefaction. The results of the parametric studies show that the subgrade reaction varies with loading rates, permeability and degrees of liquefaction, and the dimensionless subgrade reaction normalized by that without seepage has a unique relation with dimensionless time factor which consists of pile diameter, loading rate, permeability coefficient and coefficient of volume compressibility.

INTRODUCTION

Liquefaction-induced ground deformation has caused severe damages in pile foundations (e.g. Tokimatsu et al., 1996). The pseudo-static analysis with beam on Winkler foundation, which is one of the conventional seismic design methods for piles, uses ground deformation at a particular moment as an external force in order to consider the influence of ground deformation. The design method requires p-y relations, which are the relations between the subgrade reaction and the relative

---

[1] Department of Civil Engineering, Graduate School of Engineering, Tohoku University, Sendai, Japan

displacement of the pile with respect to free-field ground. The p-y relations have been investigated with the shaking table tests by many researchers [e.g., Boulanger et al. (1999), Wilson et al. (2000), and Tokimatsu et al. (2004)].

Some former studies have shown that p-y relation has apparent loading rate dependency. Experimental studies with shaking table tests have shown that the subgrade reaction correlated with the relative velocity of the pile with respect to free-field the case of loose liquefied ground [e.g. Tamura et al. (2000), Tokimatsu et al. (2001), and Suzuki and Adachi (2002)]. Moreover, physical model studies have shown that the resistance between saturated or liquefied and an embedded object (e.g. pipe, sphere, and plow) increase with the increase of velocity of the object movement [e.g. Kawakami et al. (1994), Towhata et al. (1999), and Palmer (1999)]. The apparent viscous effect in p-y relation seems to be due to the behavior of near-field soil adjacent to piles. It is however difficult to observe the behavior of near-field soil adjacent to piles through normal shaking table tests, although recent large shaking table tests make it possible to observe the near-field behavior (Tokimatsu and Suzuki, 2004). Numerical analysis is therefore more useful to observe the soil-water coupled behavior around a pile [Iai (2002) and Narita et al. (2003)].

The apparent viscous effect in p-y relation is related to the following two effects; 1) inertial effect of neighboring soil to pile (Uzuoka, 2005), 2) seepage effect of pore water pressure due to soil dilatancy [Kutter and Voss (1995), Palmer (1999), and Yoshimine (2003)]. The latter effect is discussed with numerical analysis in this study. The numerical analysis possesses the following features: 1) A simplified non-viscosity constitutive model for liquefied sand, and 2) A soil-pore water coupled formulation without inertial and viscous terms. The liquefied soil at a certain depth around a pile is carefully modeled with plane stress finite elements. The subgrade reaction of a pile in saturated or liquefied soil is calculated under various loading rates, permeability and degrees of liquefaction. The soil-water coupled behavior adjacent to a pile is discussed through the numerical results. Finally, the loading rate dependency of subgrade reaction of a single pile in saturated or liquefied soil is clarified with dimensionless parameters.

NUMERICAL METHOD

Field Equations

In this study, a soil-water coupled problem is formulated based on a u-p formulation (Oka et al., 1994). The finite element method (FEM) is used for the spatial discretization of the equilibrium equation, while the finite difference method (FDM) is used for the spatial discretization of the pore water pressure in the continuity equation. Newmark method was used for time integration. In this study, we neglect inertial and viscous terms from the field equations; therefore the field equations are similar to equations used in consolidation analysis. The governing equations are formulated with the following assumptions; 1) the infinitesimal strain, 2) the smooth distribution of porosity in the soil and 3) incompressible grain particles in the soil. The equilibrium equation for the mixture is derived as follows:

$$\frac{\partial \sigma'_{ij}}{\partial x_j} + \frac{\partial p}{\partial x_i} + \rho b_i = 0 \tag{1}$$

where σ'_{ij} = the effective stress tensor; p = pore water pressure; ρ = overall density; and b_i = body force vector. The continuity equation is derived as follows:

$$\frac{k}{\gamma_w} \nabla^2 p + \dot{\varepsilon}_{ii}^{\,s} - \frac{n}{K^f} \dot{p} = 0 \tag{2}$$

where k = coefficient of permeability; γ_w = unit weight of the fluid; $\varepsilon_{ii}^{\,s}$ = volumetric strain of the solid; n = porosity; and K^f = bulk modulus of the fluid.

Simplified Constitutive Model for Liquefied Sand

A simplified constitutive model for liquefied sand is proposed. An existing constituting model (Oka et al., 1999) is modified in order that the model can explicitly treat degree of liquefaction based on the concept of minimum effective stress. The proposed constitutive model possesses the following features.

Yield Function and Hardening Rule

The yield function is expressed as:

$$f = \left\{ \left(\eta_{ij}^* - \chi_{ij}^* \right) \left(\eta_{ij}^* - \chi_{ij}^* \right) \right\}^{1/2} - k = 0 \tag{3}$$

$$\eta_{ij}^* = s_{ij} / \sigma'_m \tag{4}$$

where σ'_m = mean effective stress; s_{ij} = deviatoric stress tensor; k = numerical parameter which defines an elastic region; and χ_{ij}^* = kinematic hardening parameter. With nonlinear kinematic hardening rule, the increment of χ_{ij}^* is given by:

$$d\chi_{ij}^* = B^* \left(M_f^* de_{ij}^P - \chi_{ij}^* d\gamma^{P*} \right) \tag{5}$$

$$d\gamma^P = \left(de_{ij}^P de_{ij}^P \right)^{1/2} \tag{6}$$

where B^* = material parameter of hardening function; M_f^* = failure stress ratio; and de_{ij}^P = plastic deviatoric incremental strain tensor.

Flow Rule and Plastic Potential Function

With non-associated flow rule the plastic potential function is expressed as:

$$g = \left\{ \left(\eta_{ij}^* - \chi_{ij}^* \right) \left(\eta_{ij}^* - \chi_{ij}^* \right) \right\}^{1/2} + \tilde{M}^* \ln \left(\sigma_m' / \sigma_{ma}' \right) = 0 \tag{7}$$

where σ_{ma}' = a constant; and $\tilde{M}^*$ is defined as follows:

$$\tilde{M}^* = \begin{cases} M_m^* & \eta^* \geq M_m^* \\ \eta^* & \eta^* < M_m^* \end{cases} \tag{8}$$

$$\eta^* = \left(\eta_{ij}^* \eta_{ij}^* \right)^{1/2} \tag{9}$$

where M_m^* = stress ratio η^* when the maximum compressive volumetric strain occurs during the shearing (phase transformation stress ratio). The liquefied soil is treated at the initial condition in this analysis; therefore plastic volumetric strain due to dilatancy is assumed to be zero when η^* is less than M_m^* as shown in equation (8).

Concept of Minimum Effective Stress

It is assumed that degree of liquefaction is related to the amount of volumetric strain with dissipation of excess pore water pressure after liquefaction. The post-liquefaction volumetric strain depends on the density of sand and the strain history [e.g. Ishihara and Yoshimine (1992), and Sento et al. (2004)]. Loose sand with much strain history yields large volumetric strain. Assuming that liquefaction process is in overconsolidation region, we can use the following stress-strain relation during the dissipation process of excess pore water pressure:

$$d\sigma_m' = \frac{(1+e_0)\sigma_m'}{\kappa} d\varepsilon_{ii}^e = \frac{1}{m_v} d\varepsilon_{ii}^e = K d\varepsilon_{ii}^e \tag{10}$$

$$\sigma_m' \geq \sigma_{ml}' = R_{\lim} \sigma_{m0}' \tag{11}$$

where $d\sigma_m'$ = incremental mean effective stress; σ_{ml}' = minimum effective stress during liquefaction; e_0 = initial void ratio; κ = swelling index; $d\varepsilon_{ii}^e$ = incremental elastic volumetric strain; m_v = coefficient of volume compressibility; K = bulk modulus; $R_{\lim}$ = ratio of σ_{ml}' for σ_{m0}' (parameter of degree of liquefaction); and σ_{m0}' = initial mean effective stress. Sento et al. (2004) changed σ_{ml}' with the density of sand and strain history, and reproduced the volumetric strain of liquefied sand with equation (10). The small σ_{ml}' means the severe degree of liquefaction. We use equation (10) for the calculation of elastic volumetric strain in the framework of elasto-plastic model. Poisson's ratio is used as another elastic coefficient. The minimum effective stress σ_{ml}' is the lower value of mean effective stress in this analysis, and is also the initial value of mean effective stress. It is noted that the σ_{ml}' has a physical meaning in this study, although the σ_{ml}' has been treated as a numerical parameter in past liquefaction analyses.

Performance of the Constitutive Model

Undrained monotonic torsional shear tests after cyclic shearing were simulated to validate the proposed constitutive model. Figure 1 shows the calculated stress strain behavior for loose Toyoura sand (the relative density of 30%) and dense Toyoura sand (the relative density of 70%). Toyoura sand is a fine uniform sand with a mean diameter D_{50} of 0.16 mm and a uniformity coefficient U_c of 1.2. The material parameters for both cases are shown in Table 1. These parameters were determined based on the results of past laboratory tests (Yoshida et al., 1994).

In the case of dense sand, the shear stress and mean effective stress recover due to the dilatancy in the strain region of over about 10 %. In the case of loose sand, the shear stress slightly recovers in the strain region of over 50 %. The effective stress path of loose sand is displayed as a point near the origin. These different behaviors

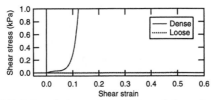

(a) Relations between shear strain and shear stress

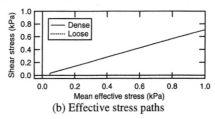

(b) Effective stress paths

Figure 1: Performance of the constitutive model

Table 1: Material parameters

Soil type		Dense sand	Loose sand
Density	ρ (t/m^3)	1.93	1.93
Initial void ratio	e_0	0.60	0.80
Swelling index	κ	0.002	0.003
Normalized initial shear modulus	G_0/σ'_{m0}	1200	900
Failure stress ratio	M^*_f	1.00	0.95
Phase transformation stress ratio	M^*_m	0.91	0.91
Hardening parameter	B^*	70	15
Initial value of σ'_m	σ'_{m0} (kPa)	49	49
Parameter for degree of liquefaction	R_{lim}	1.0×10^{-2}	1.0×10^{-8}

between loose and dense sand are mainly due to the different minimum effective stress. The minimum effective stress of dense sand is 10^6 times larger than that of loose sand. We can easily treat the change in the density and degree of liquefaction with the minimum effective stress in this model.

NUMERICAL CONDITIONS OF PARAMETRIC STUDIES

Figure 2 shows the concept of analytical model and the finite element model. The horizontal plane around a single pile at a certain depth is treated. Figure 3 shows the finite element model in the case of pile diameter of 0.5 m. Only half model in the positive of Y-axis can be used considering the symmetry. The dimensions of the finite element model are 10 m×10 m×0.05 m. The size of the model in the case of pile diameter of 1.0 m is twice, and the number of elements is the same as the case of pile diameter of 0.5 m. The soil around the pile is modeled with isoparametric 8-noded elements, and the pile is assumed to be rigid. Only one element exists in Z-direction with the height of 0.05 m.

The lateral boundaries except for symmetric surface are fixed in the horizontal directions (X and Y directions), and the bottom boundary is fixed in Z-direction. The horizontal directions of the nodes on the same position in XY plane are tied to achieve plane stress condition; therefore no shear deformation in YZ and ZX plane.

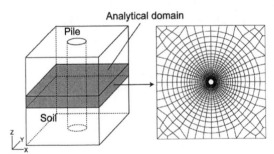

Figure 2: Concept of modeling

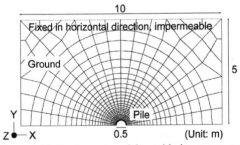

Figure 3: Finite element model considering symmetry

Table 2: Parameters for parametric studies

Parameters		Values
Diameter of pile	D (m)	0.5, 1.0
Degree of liquefaction	R_{lim}	1.0×10^{-8}, 1.0×10^{-2}, 1.0
Coefficient of permeability	k (m/s)	0.0, 1.0×10^{-4}, 1.0×10^{-2}, 1.0
Loading rate	v (m/s)	0.0005, 0.005, 0.05, 0.5, 5.0

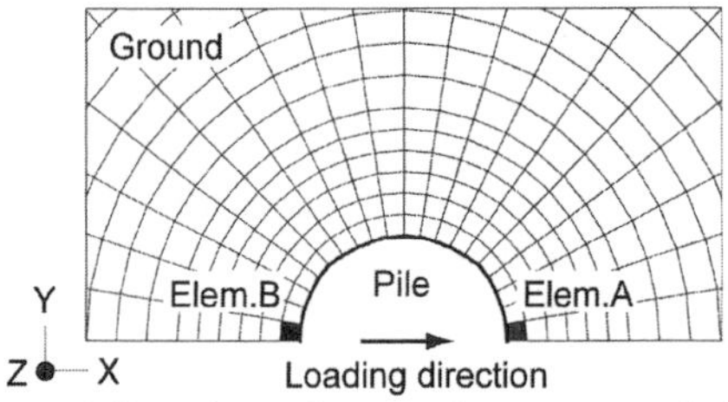

Figure 4: Locations of output elements around pile

All boundaries are impermeable. A monotonic displacement in X-direction is set at the nodes on the edge of the pile. The amplitude of displacement is half of pile diameter. The incremental displacement of the calculation is 5.0×10^{-5} m for all cases, but the displacement rate is changed for each case. The subgrade reaction can be calculated as the summation of reaction forces at nodes on the pile, and the converted value for the depth of 1.0 m will be shown in later.

All solid elements are modeled with proposed constitutive model for liquefied sand. The effective overburden pressure before liquefaction is 49 kPa, and the vertical total stress is kept constant. The initial effective stress and the excess pore water pressure are assumed for the parameter of degree of liquefaction. We consider three cases for the degree of liquefaction, severe degree ($R_{lim}=1.0\times10^{-8}$), medium degree ($R_{lim}=1.0\times10^{-2}$), and non-liquefied ($R_{lim}=1.0$) as shown in Table 2. The diameter of pile, coefficient of permeability, and loading rate are changed in the parametric studies.

NUMERICAL RESULTS OF PARAMETRIC STUDIES

Soil-Water Coupled Behavior around Pile

The results of two representative cases are discussed here. Case 1 corresponds to nearly undrained condition with $D=0.5$ m, $R_{lim}=1.0\times10^{-2}$, $k=1.0\times10^{-4}$ m/s, and $v=5.0$ m/s. Case 2 corresponds to nearly drained condition with $D=0.5$ m, $R_{lim}=1.0\times10^{-2}$, $k=1.0$ m/s, and $v=0.0005$ m/s. The locations of elements A and B for discussion are shown in Figure 4.

The time histories of stress and strain behavior at the elements in Case 1 are shown in Figure 5. The time histories show the components of a) second invariant of deviatoric strain, b) volumetric strain, c) mean effective stress, d) second invariant of deviatoric stress and subgrade reaction, e) mean total stress, and f) excess pore water pressure respectively. The lateral movement of pile causes deviatoric strain at both elements (see Figure 5a). The large deviatoric strain causes expansive plastic volumetric strain due to dilatancy; however the volumetric strain remains almost zero (see Figure 5b) because the loading rate is sufficiently faster than drainage in this case. Thereby, the mean effective stress and the deviatoric stress increase at both elements (see Figure 5c, 5d). The subgrade reaction increases with the increase of the deviatoric stress, and shows almost similar tendency with the deviatoric stress (see Figure 5d). The subgrade reaction and deviatoric stress show positive curvature due to the recovery of stiffness. In addition, the mean total stress at element B decreases behind the pile, while the mean total stress at element A increases in front of the pile (see Figure 5e). Thereby, element B shows larger amount of decrease in the excess pore water pressure than element A, although the excess pore water pressure decreases at both elements (see Figure 5f).

The time histories of stress and strain behavior at the elements in Case 2 are also shown in Figure 6. The components are the same as those in Figure 5. The lateral movement of pile causes larger deviatoric strain at both elements than that in Case 1 (see Figure 6a). In particular, extreme large deviatoric strain is generated at the element B behind the pile. The large deviatoric strain causes expansive volumetric

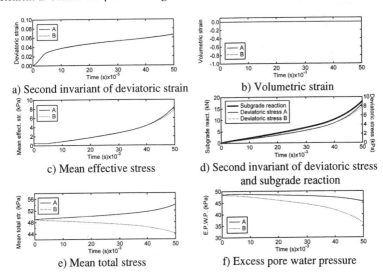

a) Second invariant of deviatoric strain

b) Volumetric strain

c) Mean effective stress

d) Second invariant of deviatoric stress and subgrade reaction

e) Mean total stress

f) Excess pore water pressure

Figure 5: Time histories of stress and strain in the Case 1

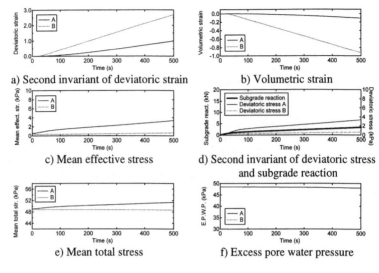

a) Second invariant of deviatoric strain

b) Volumetric strain

c) Mean effective stress

d) Second invariant of deviatoric stress and subgrade reaction

e) Mean total stress

f) Excess pore water pressure

Figure 6: Time histories of stress and strain in the Case 2

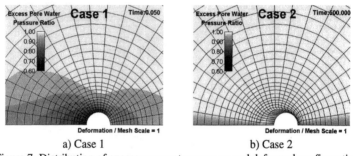

a) Case 1   b) Case 2

Figure 7: Distribution of excess pore water pressure and deformed configuration

strain due to dilatancy (see Figure 6b) because the loading rate is sufficiently slower than drainage in this case. Thereby, the mean effective stress and the deviatoric stress increase slower than Case 1 (see Figure 6c, 6d). The subgrade reaction increases with the increase of the deviatoric stress, and shows almost similar tendency with the deviatoric stress (see Figure 6d). Contrary to Case 1, the subgrade reaction and deviatoric stress do not show positive curvature because the recovery of stiffness is less than Case 1. As a result, the subgrade reaction is less than that in Case 1. In addition, the mean total stress at element B decreases behind the pile, while the mean total stress at element A increases in front of the pile (see Figure 6e). Thereby, element A shows larger amount of increase in the mean effective stress than element

B, because the excess pore water pressure is not produced due to drainage (see Figure 6f).

Figure 7a and 7b show the final distributions of excess pore water pressure ratio and final deformed configurations of Case 1 and 2 respectively. The displacement of the pile is 0.25 m at the end of the calculation. The excess pore water pressure ratio means the ratio of excess pore water pressure to the initial overburden effective stress. The initial excess pore water pressure ratio is almost 1.0 in both cases. The decrease of excess pore water pressure is depicted behind the pile in Case 1 (see Figure 7a) as shown in Figure 5f. The deformed area around the pile spreads widely. Contrary to Case 1, no change in excess pore water pressure is shown in Case 2 (see Figure 7b), and the deformation is localized adjacent to the pile.

Relation between Normalized Subgrade Reaction and Time Factor

The parametric studies show that the subgrade reaction varies with loading frequencies, permeability and degrees of liquefaction. In this section we introduce some dimensionless parameters to discuss the rate-dependency of subgrade reaction. Figure 8 shows the relations between dimensionless subgrade reaction and dimensionless permeability for the cases of pile diameter of 0.5 m. The dimensionless subgrade reaction P^* is defined as:

$$P^* = P / P_{ud} \tag{12}$$

where P = subgrade reaction; and P_{ud} = subgrade reaction without seepage. The subgrade reactions when the displacement of pile is 0.25 m are shown in Figure 8. The dimensionless permeability k^* is defined as:

$$k^* = k / v \tag{13}$$

where v = velocity of pile. The relation between dimensionless subgrade reaction and dimensionless permeability can be represented as a unique line for each degree of liquefaction. If the degree of liquefaction becomes severe, the dimensionless subgrade reaction becomes close to 1.0. This is due to less effect of dilatancy of severely liquefied soil.

We introduce time factor used in consolidation analysis in order to consider the degree of liquefaction. The time factor T^* is defined as:

$$T^* = \frac{k}{\gamma_w m_v D v} = \frac{kK}{\gamma_w D v} \tag{14}$$

where D = diameter of pile. It is noted that the coefficient of volume compressibility m_v or the bulk modulus K are the initial values in equation (14). Figure 9 shows the relations between dimensionless subgrade reaction and time factor for all cases. The dimensionless subgrade reaction has a unique relation with time factor which consists of pile diameter, loading rate, permeability coefficient and

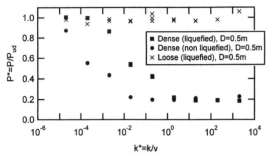

Figure 8: Relations between dimensionless subgrade reaction and dimensionless
permeability for the cases of pile diameter of 0.5 m

coefficient of volume compressibility. It is noteworthy that the rate-dependency of
subgrade reaction can be characterized with time factor; however, the relations are
dependent on the displacement of pile. Figure 10 shows the relations for different
pile displacements of 0.25 m and 0.125 m. If the displacement of pile becomes small,
the relation becomes close to 1.0 because of less effect of dilatancy. The dilatancy
behavior is essentially nonlinear; therefore the relations are dependent on the
displacement. We need further study on the relations considering the magnitude of
displacement.

Figure 10 shows that the rate-dependency of subgrade reaction becomes clear
under large deformation, and vanishes in loose liquefied soil. This tendency seems to
be contrary to some past results of shaking table tests. Some experimental studies
with shaking table tests have shown that the subgrade reaction is correlated with the
relative velocity of the pile with respect to free-field in case of loose liquefied ground
[e.g. Tamura et al. (2000), Suzuki and Adachi (2002)]. This discrepancy can be
explained with another rate-effect, namely inertial effect of neighboring soil to pile
(Uzuoka, 2005).

CONCLUSIONS

The loading rate dependency of the subgrade reaction of a pile in saturated or
liquefied soil is discussed through numerical analyses. A simplified non-viscous
constitutive model for sand is incorporated with the soil-water coupled formulation.

The numerical results reveal the following:
1) The subgrade reaction varies with loading frequencies, permeability and degrees
   of liquefaction.
2) The subgrade reaction becomes large due to much dilatancy in the case with fast
   loading rate, low permeability and soft degree of liquefaction. The decrease in the
   excess pore water pressure ratio is remarkable behind a pile.
3) The subgrade reaction becomes small due to little dilatancy in the case with slow

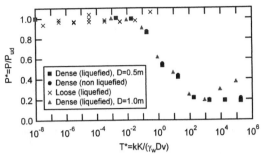

Figure 9: Relations between dimensionless subgrade reaction and time factor for all cases (displacement of pile: 0.25 m)

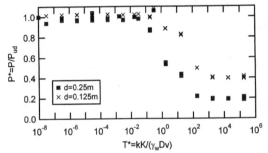

Figure 10: Influence of displacement of pile on the relations between dimensionless subgrade reaction and time factor

loading rate, high permeability and severe degree of liquefaction. The deformation of soil is localized adjacent to a pile.

4) The dimensionless subgrade reaction has a unique relation with dimensionless time factor. The rate-dependency of subgrade reaction can be characterized with time factor; however, the relations are dependent on the displacement of pile.

In this study, the horizontal plane was the only object under consideration; however, the responses of liquefied ground and pile also change along the vertical direction. We require further investigations into the 3-dimensional behavior of piles and ground.

ACKNOWLEDGEMENTS

This study was supported by the research grant of Japan Iron and Steel Federation in FY2002-2004. Miss Minako Shibasaki, formerly graduate student of Tohoku University, carried out the numerical analyses. The author wishes to thank them for their cooperation.

REFERENCES

Boulanger, R. W., Curras, C. J., Kutter, B. L., Wilson, D. W., and Abghari, A. (1999). "Seismic soil-pile-structure interaction experiments and analyses," *Journal of Geotechnical and Geoenvironmental Engineering*, ASCE, 125(9), 750-759.

Iai, S. (2002). "Analysis of soil deformation around a cylindrical rigid body," *Proc. of the U.S.-Japan Seminar on Seismic Disaster Mitigation in Urban Area by Geotechnical Engineering*, Alaska, 739-746.

Ishihara, K. and Yoshimine, M. (1992). "Evaluation of settlements in sand deposits following earthquakes," *Soils and Foundations*, 32(1), 173-188.

Kawakami, T., Suemasa, N., Hamada, M., Sato, H., and Katada, T. (1994). "Experimental study on mechanical properties of liquefied sand," *Proceedings from the 5th U.S.-Japan Workshop on Earthquake Resistant Design of Lifeline Facilities and Countermeasures Against Soil Liquefaction*, Technical Report NCEER-94-0026, M. Hamada and T. D. O'Rourke, eds., State University of New York, Buffalo, 285-299.

Kutter, B. L., and Voss, T. (1995). "Analysis of data on plow resistance in dense, saturated, cohesionless soil," *Contract Report.* CR95.004, Naval Facilities Engineering Service Center, Prot Hueneme, California.

Narita, N., Sako, H., and Tokimatsu, K. (2003). "Effect of permeability on interaction between soil and pile during earthquake, *Proc. 38th Japan National Conf. on Geotechnical Engineering*, Akita, 1903-1904. (in Japanese)

Oka, F., Yashima, A., Shibata, T., Kato, M. and Uzuoka, R. (1994). "FEM-FDM coupled liquefaction analysis of a porous soil using an elasto-plastic model," *Applied Scientific Research*, 52, 209-245.

Oka, F., Yashima, A., Tateishi, A., Taguchi, Y. and Yamashita, S. (1999). "A cyclic elasto-plastic constitutive model for sand considering a plastic strain dependency of the shear modulus," *Geotechnique*, 49(5), 661-680.

Palmer, A. C. (1999). "Speed effects in cutting and ploughing," *Geotechnique*, 49(3), 285-294.

Sento, N., Kazama, M. and Uzuoka, R. (2004). "Experiment and idealization of the volumetric compression characteristics of clean sand after undrained cyclic shear," *J. Geotechnical Engineering*, JSCE, 764/III-67, 307-317. (in Japanese)

Suzuki, Y., and Adachi, N. (2003). "Relation between subgrade reaction of pile and liquefied ground response," *Proc. 38th Japan National Conf. on Geotechnical Engineering*, Akita, 1945-1946. (in Japanese)

Tamura, S., Kobayashi, K., Suzuki, Y., and Yoshizawa, M. (2001). "Relation between pile behavior and subgrade reaction based on pile vibration test using large-scale laminar box," *Proc. 36th Japan National Conf. on Geotechnical Engineering*, Tokushima, 1711-1712. (in Japanese)

Tokimatsu, K., Mizuno, H., and Kakurai, M. (1996). "Building damage associated with geotechnical problems," *Soils and Foundations*, Special Issue on Geotechnical Aspects of the January 17 1995 Hyogoken-Nambu Earthquake, No.1, 219-234.

Tokimatsu, K., Suzuki, H., and Suzuki, Y. (2001). "Back-calculated p-y relations of liquefied soils from large shaking table tests," *Proc of 4th International*

Conference on Recent Advances in Geotechnical Earthquake Engineering and Soil Dynamics, Prakash, S. ed., San Diego, Paper No.6.24.

Tokimatsu, K., Suzuki, H., and Sato, M. (2004). "Influence of inertial and kinematic components on pile response during earthquake," Proc. of 11th International Conference on Soil Dynamics & Earthquake Engineering and 3rd International Conference on Earthquake Geotechnical Engineering, Berkeley, 768-775.

Tokimatsu, K. and Suzuki, H. (2004). "Pore water pressure response around pile and its effects on p-y behavior during soil liquefaction," Soils and Foundations, 44(6), 101-110.

Towhata, I., Vargas-Mongeb, W., Orensec, R. P., and Yaod, M. (1999). "Shaking table tests on subgrade reaction of pipe embedded in sandy liquefied subsoil," Soil Dynamics and Earthquake Engineering, 18, 347–361.

Uzuoka, R. (2005). "Loading rate dependency of the subgrade reaction for a pile in liquefied ground," Proc. of 16th International Conference on Soil Mechanics and Geotechnical Engineering, Osaka. (in press)

Wilson, D. W., Boulanger, R. W., and Kutter, B. L. (2000). "Observed seismic lateral resistance of liquefying sand," Journal of Geotechnical and Geoenvironmental Engineering, ASCE, 126(10), 898-906.

Yoshida, N., Yasuda, S., Kiku, M., Masuda, T. and Finn, W. D. L. (1994). "Behavior of sand after liquefaction," Proc. from the 5th U.S.-Japan Workshop on Earthquake Resistant Design of Lifeline Facilities and Countermeasures Against Soil Liquefaction, Technical Report NCEER-94-0026, M. Hamada and T. D. O'Rourke, eds., State University of New York, Buffalo, 181-198.

Yoshimine, M. (2003). "Liquefied soil-pile interaction and its rate effects," First Japan-U.S. Workshop on Testing, Modeling, and Simulation in Geomechanics, Boston. (in press)

DYNAMIC ANALYSES OF SOIL-PILE-STRUCTURE INTERACTION IN LATERALLY SPREADING GROUND DURING EARTHQUAKE SHAKING

Dongdong Chang, Student Member ASCE [1], Ross W. Boulanger, Member ASCE [2], Scott J. Brandenberg, Member ASCE [1], and Bruce L. Kutter, Member ASCE [2]

ABSTRACT

Nonlinear dynamic analyses using the finite element (FE) method are compared to the recorded responses of pile-supported structures in soil profiles that develop liquefaction and lateral spreading during earthquake shaking in centrifuge model tests. The centrifuge tests included a simple superstructure supported on a group of six piles. The soil profile consisted of a gently sloping nonliquefiable clay crust over liquefiable loose sand over dense sand. The FE models consisted of a two-dimensional model of the soil profile with a beam-column model of the structure and pile foundation that was attached to the soil profile by soil springs. Centrifuge recordings were used to calibrate the nonlinear FE models. Parametric analyses were used to evaluate the sensitivity of the FE results to variations in the modeling parameters and FE representation. With appropriate calibrations, the FE models were able to reasonably approximate the essential features of soil and structural response.

INTRODUCTION

A major cause of damage to bridges in past earthquakes has been liquefaction and associated lateral spreading of the foundation soils. Research in recent years has led to significant advances in our understanding of the mechanics of soil-pile-structure interaction in liquefying and laterally spreading ground (e.g. see reviews by Dobry

---

[1] Graduate Student Researcher, Department of Civil & Environmental Engineering, University of California, Davis, CA, 95616.
[2] Professor, Department of Civil & Environmental Engineering, University of California, Davis, CA, 95616.

and Abdoun 2003, Tokimatsu 2003, Boulanger et al. 2003), but there continues to be numerous questions and issues that need to be resolved for advancing pseudo-static and dynamic analysis methods.

This paper presents comparisons of nonlinear dynamic finite element (FE) analyses and dynamic centrifuge model tests of simple superstructures supported on pile groups founded in soil profiles that develop liquefaction and lateral spreading during earthquake shaking. Details of the centrifuge models and the finite element models are briefly summarized. Then the results of the FE and centrifuge models are compared in terms of time series of soil and structural responses, and the magnitude and timing of the lateral spreading and inertial loads throughout shaking. The results show that the FE models were able to reasonably approximate the essential features of soil and structural response for the range of conditions covered in the experiments (eight cases). After the baseline model parameters had been appropriately calibrated, parametric analyses were used to evaluate the sensitivity of the FE results to variations in the modeling parameters and FE representation. Further parametric studies will include a broader range of geotechnical and structural conditions. Insights regarding the mechanisms of soil-pile-foundation-structure interaction during liquefaction and lateral spreading are summarized, and the implications for design using simplified pseudo-static pushover analysis are discussed.

CENTRIFUGE MODELS

A schematic cross-section of one of the centrifuge models is shown in Figure 1. The soil profile consisted of a non-liquefied crust overlying loose sand ($D_r \approx 35\%$), which in turn was overlying dense sand ($D_r \approx 75\%$). The crust layer sloped gently (3°) toward a channel carved in the crust at one end of the model.   The non-liquefiable crust consisted of reconstituted Bay Mud with undrained shear strength of about 30 kPa in one model (named DDC01) and 20 kPa in another model (named DDC02). A thin layer of coarse Monterey sand was placed on the surface of the Bay Mud to prevent it from drying during centrifuge spinning.

The simple superstructure was supported on a six-pile group that consisted of 1.17-m diameter piles. Dimensions are given in prototype terms unless otherwise indicated. The pile cap provided essentially fixed-head restraints for the piles. The superstructures had a prototype fixed-base natural period of 0.8s in DDC01 and 0.3s in DDC02.

Each test was shaken with a number of simulated earthquakes conducted in series with sufficient time between shakes to allow dissipation of excess pore pressures. Each shaking event was stronger than the previous one, with the first shakes having $a_{max,base}$ of about 0.15g and the last shakes having $a_{max,base}$ of about 0.65g.

Details of these centrifuge experiments are summarized in a series of data reports (Chang et al. 2004a, b) available from the web site for the Center for Geotechnical Modeling at UCD (http://cgm.engr.ucdavis.edu). These data reports include detailed explanations of model construction, data acquisition procedures, data organizational structure, and post-earthquake model dissection measurements.

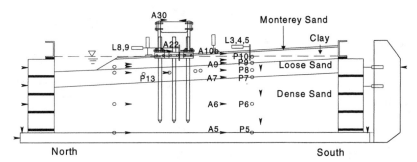

Figure 1. Cross section of centrifuge model DDC02 showing the instrumentation.

FE MODELS

The FE modeling was performed using OpenSees (http://opensees.berkeley.edu/) as follows. The soil continuum was modeled using 2D quad elements with Yang et al. (2003) mixed stress-strain space, pressure-dependent, multiple yield surface constitutive model for the sands and the pressure-independent, multiple yield surface consitutive model for the clay. The model parameters describing the maximum (low-strain) shear modulus were determined from in-flight shear wave velocity measurements, and then the model parameters for the various yield surfaces are set based on the soil's specified G/G_{max} relation and ultimate shear strength parameters. The parameters controlling liquefaction behavior were then specified to produce a desired cyclic resistance ratio (CRR) and limiting cyclic shear strains.

The lateral (p-y) and axial (shaft t-z, tip q-z) soil springs were modeled using the material models developed by Boulanger et al. (2004). These soil spring materials are coupled to Yang et al.'s (2003) soil models to account for excess pore water pressure generation in the adjoining soils. The soil springs are modeled as zero-length elements that generally share a node with adjoining solid soil elements. Specifying two solid soil elements allows the p-y material to depend on effective stresses (and hence excess pore pressures) above and below its nodal position (essentially covering its tributary length). Additional constraints on the constitutive response, including a residual capacity, are described in Boulanger et al. (2003). The parameters controlling the soil springs along the piles were specified based on API (1993) recommendations for sands and clays.

Lateral springs were also used to model the total crust load against the pile cap. These springs were set to approximate the crust load-transfer relation that was developed by Brandenberg et al. (2004).

Interface elements were placed between the clay crust and the top of the loose sand layer. These interface elements were one dimensional elasto-plastic zero length elements with a varying strength scaled by the excess pore water pressure of the underlying loose sand layer, schematically shown in Figure 2. The residual shear strength of the interface elements, once the underlying soil had liquefied, was set at 3 kPa because it produced lateral spreading deformations consistent with the measured displacement discontinuity between the clay crust and the liquefied sand layer.

Finite element pre- and post- processing software GiD was used as an interface for OpenSees input and output. The two-dimensional FE mesh (generated in GiD) for one of the centrifuge models DDC02 is shown in Figure 3, where 3-(a) is the undeformed mesh before earthquake shaking, and 3-(b) is the deformed mesh without interface elements after a large Kobe base motion, and 3-(c) is the deformed mesh with interface elements after the same large Kobe base motion. The latter mesh simulated the displacement discontinuity that was observed between top clay crust and underlying loose sand layer in the centrifuge tests.

Boundary conditions for the two-dimensional FE models due to the constraints of the flexible container were treated using two different methods: the first method was to equate the lateral degree of freedom for nodes at the same elevations along the container wall at the two ends of the model, and the second method was to simulate the flexible container in the FE mesh. Undeformed and deformed FE meshes for this two methods are shown in Figure 4. Comparisons of FE analysis with and without the container in the FE mesh are still being evaluated. The FE analysis results presented in this paper are for the first method (as shown in Figure 4-b).

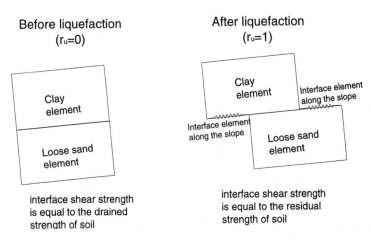

Figure 2. Schematics of zero length interface elements between top nonliquefiable clay crust and underlying liquefiable sand layer.

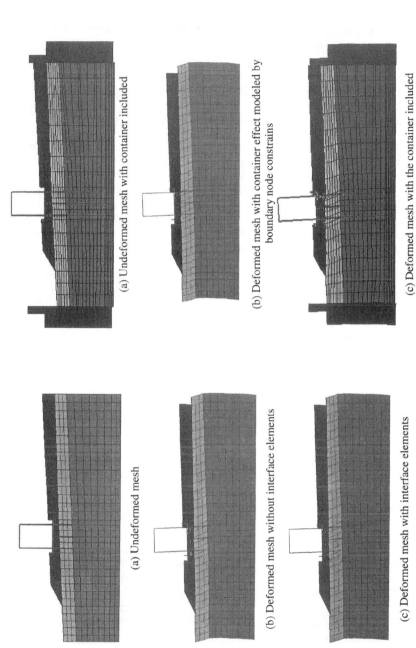

(a) Undeformed mesh with container included

(b) Deformed mesh with container effect modeled by boundary node constrains

(c) Deformed mesh with the container included

Figure 4. FE mesh showing undeformed and deformed geometries

(a) Undeformed mesh

(b) Deformed mesh without interface elements

(c) Deformed mesh with interface elements

Figure 3. FE mesh showing undeformed and deformed

COMPARISON OF CALCULATED AND RECORDED RESPONSES

The challenge for the FE models was to reasonably approximate the recorded responses for a total of eight cases (two models, each shaken by four different earthquakes) using a common set of FE modeling parameters. For a given "set" of FE modeling parameters, the only parameters that were changed between FE models of the two centrifuge tests was the superstructure column stiffness (sets the fixed based period), the undrained shear strength of the clay layer (set to match the measured values), and the layer thicknesses (also set to match the as-built geometries).

Typical Time Series of Model DDC02 During a Large Kobe Motion

Figures 5 and 6 compare the recorded and calculated accelerations in the dense sand, loose sand, clay crust, pile cap, and superstructure for the DDC02 model during a large Kobe motion. The FE analysis reproduced structural accelerations reasonably well, and captured the low frequency trends of soil accelerations but missed the high frequency spikes.

Figures 7 and 8 compare the bending moments and shear forces at a pile head, and the cap and clay crust lateral displacements, together with the excess pore water pressure ratios r_u in the dense and loose sand. The FE analysis reasonably reproduced the bending moment and shear at the pile head, but over-predicted the dynamic cap displacements and under-predicted the dynamic crust displacements. The FE analysis captured some of the dilation spikes of r_u in the loose sand layer, but missed most of them in the dense sand. Missing the dilation spikes in the liquefied sand layer may affect the calculated peak loads since peak inertia and crust load tend to occur at the temporary stiffening of the liquefied soil at the dilation spikes.

Figures 9 and 10 compare the recorded total shear, crust load, and inertial forces with the FE computed lateral loads, together with the relative displacements between pile cap and clay crust. The FE analysis captured the essential trends of the critical loading conditions and phasing of lateral loads, which shows that peak cap inertia and peak superstructure inertia tend to be in phase, together with a large fraction of the peak crust load that also coincides with the peak inertia loads (Chang et al. 2005).

The soil and structural accelerations are presented in terms of acceleration response spectra in Figure 11, which shows that the FE analysis reasonably captured the low frequency content of the accelerations. The response spectra show that accelerations in the clay layer are significantly different with those in the dense sand layers since liquefaction of the loose sand amplified the low frequency components and de-amplified the high frequency components of the base motions. The results of FE analysis could capture those trends appropriately.

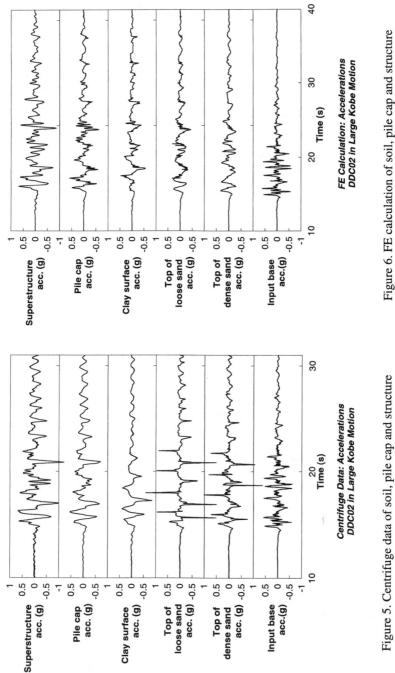

Figure 6. FE calculation of soil, pile cap and structure accelerations.

Figure 5. Centrifuge data of soil, pile cap and structure accelerations.

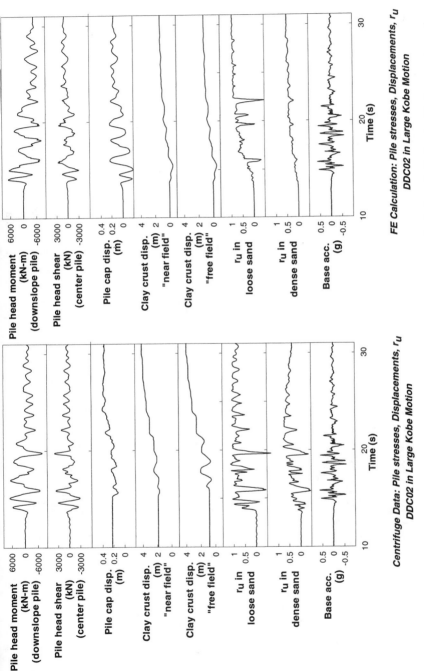

Figure 8. FE calculation of pile loads, displacements, and r_u.

Figure 7. Centrifuge data of pile loads, displacements, and r_u.

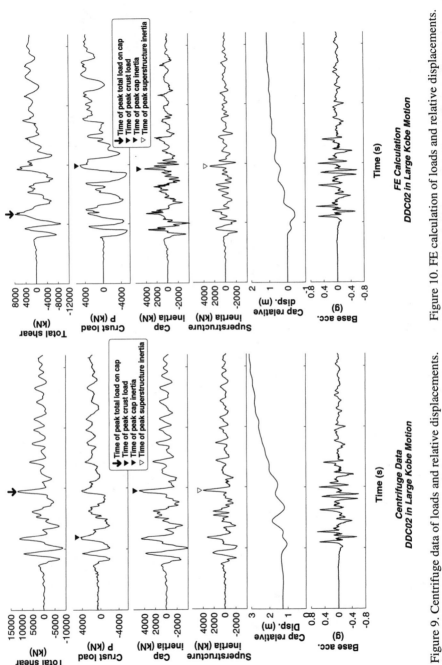

Figure 10. FE calculation of loads and relative displacements.

Figure 9. Centrifuge data of loads and relative displacements.

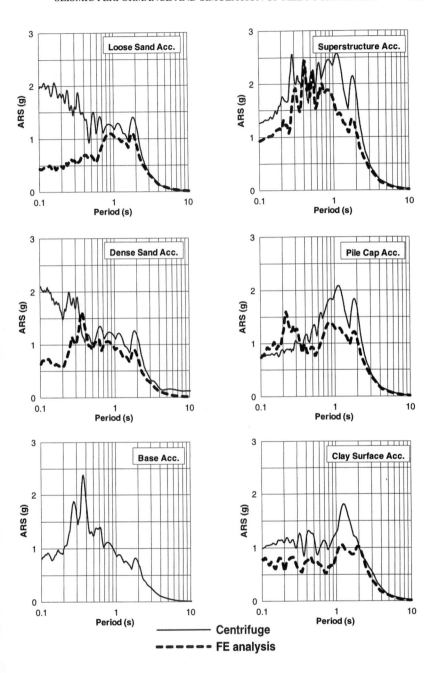

Figure 11. Comparison of soil, pile and structure acceleration response spectra between FE calculations and centrifuge data.

Effect of FE model details

Sensitivities of the two dimensional nonlinear dynamic OpenSees analysis results to various FE model input parameters were evaluated. For example, analyses were repeated for the two methods modeling container (e.g. Figure 4), for cases with different FE mesh size, and for cases with and without the interface slip elements beneath the clay layer. Other influencing factors for the FE analysis such as the pile cap load-transfer relation, residual strength of interface elements, undrained shear strength of the clay crust, thickness of liquefiable loose sand layer, relative density of different sand layers, foundation rigidity, natural periods of the superstructure, shaking level and frequency content of the input base motions, still need to be studied systematically. Greater details of the sensitivity study of the FE analysis results to input parameters will be presented in future papers.

SUMMARY

The nonlinear dynamic FE analyses were reasonably able to approximate the recorded responses of pile-supported structures in soil profiles that developed liquefaction and lateral spreading during earthquake shaking in the two centrifuge tests. The challenge was to approximate the recorded responses for a total of eight cases (two models, each shaken by four different earthquakes) using a common set of FE modeling parameters. As illustrated by the representative time series presented herein, the FE models do have some limitations in capturing all measurements of soil and structural responses equally well. Nonetheless, the overall comparisons indicated that these calibrated FE models could now be used to parametrically evaluate the influence of other key factors, such as varying structural periods, pile configurations, soil profiles, and input motions.

ACKNOWLEDGMENTS

Funding was provided by Pacific Earthquake Engineering Research (PEER) Center, through the Lifeline Program and the Earthquake Engineering Research Centers Program of the National Science Foundation, under contract 2312001. The centrifuge tests were funded by Caltrans under contract numbers 59A0162 and 59A0392. The contents of this paper do not necessarily represent a policy of either agency or endorsement by the state or federal government. Recent upgrades to the centrifuge have been funded by NSF award CMS-0086566 through the George E. Brown, Jr. Network for Earthquake Engineering Simulation (NEES). The authors appreciate the assistance that Zhaohui Yang and Ahmed Elgamal generously provided for their constitutive models.

REFERENCES

API (1993). *Recommended Practice for Planning, Design, and Constructing Fixed Offshore Platforms.* API RP 2A - WSD, 20th ed., American Petroleum Institute.

Boulanger, R. W., Kutter, B. L., Brandenberg, S. J., Singh, P., and Chang, D. (2003). "Pile foundations in liquefied and laterally spreading ground during earthquakes: Centrifuge experiments and analyses." Report CGM/2003-01, Center for Geotechnical Modeling, University of California, Davis, CA, 205 pp.

Boulanger, R. W., Wilson, D. W., Kutter, B. L., Brandenberg, S. J., and Chang, D. (2004). "Nonlinear FEM analyses of soil-pile interaction in liquefying sand." Proceedings, Geotechnical Engineering for Transportation Projects, Geotechnical Special Publication No. 126, ASCE, 470-478.

Brandenberg, S., Boulanger, R., Kutter, B., Wilson, D., and Chang, D. (2004). "Load transfer between pile groups and laterally spreading ground during earthquakes." *13th World Conference on Earthquake Engineering Vancouver,* University of California at Davis, paper No. 1516.

Chang, D., Boulanger, R. W., Kutter, B. L, and Brandenberg, S. J. (2005). "Experimental observations of inertial and lateral spreading loads on pile groups during earthquakes." Proceedings, GeoFrontiers Conference, Austin, Texas, Geotechnical Special Publication, ASCE.

Chang, D., Brandenberg, S. J., Boulanger, R. W., and Kutter, B. L. (2004a). "Behavior of piles in laterally spreading ground during earthquakes – centrifuge data report for DDC01." Report No. UCD/CGMDR-04/01, Center for Geotechnical Modeling, Department of Civil Engineering, University of California, Davis.

Chang, D., Brandenberg, S. J., Boulanger, R. W., and Kutter, B. L. (2004b). "Behavior of piles in laterally spreading ground during earthquakes – centrifuge data report for DDC02." Report No. UCD/CGMDR-04/02, Center for Geotechnical Modeling, Department of Civil Engineering, University of California, Davis.

Dobry, R., Abdoun, T., O'Rourke, T.D., and Goh, S.H. (2003). "Single Piles in Lateral Spreads: Field Bending Moment Evaluation." *J. of Geotechnical & Geoenvironmental Engrg.,* ASCE, Vol. 129(10), 879-889.

Tokimatsu, K. (2003). "Behavior and design of pile foundations subjected to earthquakes." Proc., *12th Asian Regional Conference on Soil Mechanics and Geotechnical Engineering,* Singapore, August 4th – 8th.

Yang, Z., Elgamal, A., and Parra, E. (2003). "Computational model for cyclic mobility and associated shear deformation." *J. Geot. and Geoenv. Engrg.,* ASCE, 129(12), 1119-1127.

BEAM ON WINKLER FOUNDATION METHOD FOR PILES IN LATERALLY SPREADING SOILS

Akihiro Takahashi, M.ASCE[1], Hideki Sugita[1] and Shunsuke Tanimoto[1]

ABSTRACT

The paper presents application of beam on non-linear Winkler foundation method to piles in laterally spreading liquefied soils. In many cases, parameters used in the method are determined to fit particular case studies and are hardly applicable to the other separate cases, due to difficulties in the modelling of complicated phenomena with the simplified method. In this study, variation of p-y curve parameters for piles in laterally spreading liquefied soils is systematically examined by numerically analysing physical model tests undertaken by independent researchers.

INTRODUCTION

One of the major sources of earthquake-induced damage to pile foundations is lateral spreading of liquefied soils. In practice, to assess performance of piles subjected to kinematic loading due to large ground deformation, the beam on non-linear Winkler foundation method is used in simplified design procedures and has been introduced in some of practical design codes in Japan (JSCE 2000). This involves pseudo-static analysis of a pile foundation subjected to a superstructure inertial force (if applicable) and soil movement through Winkler springs that model soil-pile interaction. When this technique is applied to a problem on a foundation in laterally spreading soils, accumulation of displacements in cyclic loading may be modelled in 'monotonic' way, which involves considerable uncertainties and makes it difficult to determine appropriate parameters for non-linear Winkler springs (p-y curves).

---

[1] Public Works Research Institute, Tsukuba, Japan.

In many cases, determination of parameters used in the method are made using the limited number of physical model test simulations or case studies and are hardly applicable to the other separate cases (e.g. Brandenberg *et al.*, 2001), due to difficulties in the modelling of complicated phenomena with the simplified method. For instance, effects of cyclic loading and nature of input earthquake motion cannot be explicitly considered in the monotonic loading numerical analysis. This kind of effects in the monotonic loading analysis has not been fully understood yet and is hidden in the model parameters.

To assess sensitivity of parameters for the method on foundation responses in laterally spreading liquefied soils, many case studies are required. For such purpose, both field observations and physical model tests may be utilised with careful regard to their advantages and limitations: Data themselves are very useful but always involve uncertainty in cause of events for the former, while foundation responses during an earthquake are clear but the tests are always conducted under unpreferable side boundary conditions for the latter. When prediction methods for ground deformation are assessed, the former may be preferable, while it is the latter for the non-linear *p-y* curve parameters. In this study, variation of *p-y* curve parameters for piles in laterally spreading liquefied soils is systematically examined by numerically analysing physical model tests undertaken by independent researchers using the beam on non-linear Winkler foundation method with hyperbolic type *p-y* curves.

PHYSICAL MODEL TEST RESULTS USED

The target situation is that a pile foundation located in loose sand deposit behind gravity type quay wall moves seaward during earthquake due to lateral liquefied soil spreading initiated by quay instability. Schematic view of the problem to be solved is illustrated in Fig. 1. Data of physical model tests on the target problem were collected from well-documented technical papers and reports.

Summary of the physical model tests used are listed in Tables 1 and 2 in ascending order of the model pile diameter. Data of six different series of physical model tests are used: Half of them are ordinary shaking table tests and the others are dynamic centrifuge tests. (All the values for the centrifuge tests are presented in the prototype scale.) The selected physical models have various configurations of soil layers (H_{NL} & H_L) and distance between foundation and quay wall (s). Test numbers 1 to 17 are the tests on single pile and the last test number 18 is on a pile group. The pile foundation in No. 18 consists of four piles rigidly fixed to the pile cap having a pile spacing of $2.5D$. All the piles used in the physical model tests do not reach their elastic limits and behave as elastic beams.

NUMERICAL PROCEDURE

The beam on non-linear Winkler foundation method with hyperbolic type *p-y* curves is used. The vertical beam models the pile and the horizontal springs connecting the beam and supporting ground model soils. The governing differential equation for the model can be expressed as

$$EI\frac{d^4y}{dz^4} = -Dp$$

and the p-y springs modelled by a hyperbolic function can be written as

$$p = C_i \frac{k_{hr}}{1+|y/y_r|} y$$

where EI=flexural rigidity of pile, y=relative displacement between pile (u) and soils in free field (u_g), z=depth from the pile head, D=width of pile (or width of pile cap, B), p=horizontal subgrade reaction, C_i=scaling factor for the p-y curve at i-th layer, k_{hr}=coefficient of initial subgrade reaction parameter, and y_r=reference relative displacement. As $p|_{y=\infty} = k_{hr}y_r$ when C_i=1, y_r may be defined as follow using the Broms's ultimate pile resistance (1964):

$$y_r = 3K_P\sigma'_v/k_{hr}$$

where K_P=coefficient of passive earth pressure, and σ'_v=effective overburden pressure.

In order to systematically determine the soil parameters for various test conditions, the following assumptions are made:

- ϕ' for all the sands used is assumed as 40 degrees, as materials used in the physical model tests are mainly clean sand. As a result, K_P=4.6 for all the cases.
- Shear modulus of sands is assumed to be expressed as $G_0 = 75F(e)(\sigma'_v/p_a)^{0.40}$ where G_0=shear modulus at very small strain (in MPa), e=void ratio, $F(e)=(2.17-e)^2/(1+e)$, and p_a=atmosphere pressure. This relationship is obtained from bender element tests on Toyoura sand in a triaxial cell (Takahashi, 2004). By assuming v=0.3, Young's modulus, $E_0 = 2(1+v)G_0$.
- k_{hr} is estimated by (JRA, 2002):

$$k_{hr} = \alpha_i \frac{E_s}{0.3}\left(\frac{B_E}{0.3}\right)^{-3/4} = \alpha_i k_{h0}$$

where α_i=additional scaling factor for the coefficient of initial subgrade reaction parameter at i-th layer, E_s=soil's Young's modulus, and B_E=effective width of foundation (=B (width of foundation) for pile cap, and =$\sqrt{D/\beta}$ for pile where β= stiffness ratio of soil to pile (=$\sqrt[4]{\overline{k}_{h0}D/4EI}$, in 1/m)). $\overline{k}_{h0}$ for β is the average value of k_{h0} from the depth of pile head to $1/\beta$. Since it is common to choose k_h at 10^{-2}m order of displacement (strain level of 10^{-2}) as the coefficients of the initial subgrade reaction in the practical pile design procedure, E_s is assumed to be $E_0/10$.

Since the analyses presented here are performed at a snapshot in time, selection of the target time in an earthquake is crucial. In this study, the time just after shaking is selected and the measured soil displacement profile at the time is input, since (1) our main concern in this paper is assessment of p-y curve parameters for predicting permanent foundation deformation due to liquefaction-induced lateral spreading of soils, and (2) the horizontal pile displacement monotonically increases with shaking and ceases to increase at the end of shaking for all the physical model tests. (No superstructure inertial force is considered.) The other boundary conditions such as constraint conditions of the beam-ends corresponding to the pile head and pile tip are set accordingly.

In the above equations, C_i and α_i are fitting parameters: as $p_{max} = C_i 3K_P \sigma'_v$ and $\frac{dp}{dy}\Big|_{y=0} = \alpha_i C_i k_{h0}$, C_i can be used for scaling of overall p-y curve and α_i can give additional reduction to the initial subgrade reaction. These may be governed by magnitude of accumulated excess pore water pressure for the liquefiable layer and probably loosening of soil for the surface non-liquefiable layer as well as effects of accumulation of displacements in cyclic loading in the monotonic loading analysis.

C_i and α_i are estimated by minimising following residual error (ΔE):

$$\Delta E = \sqrt{\sum_{i=1}^{N}\left(\frac{M_i^* - M_i}{M_i}\right)^2 w_i}$$

where N=number of bending moment measurement points, M_i^*=estimated bending moment at i-th measurement point, M_i=measured bending moment, and $w_i = M_i^2/\Sigma M_i^2$ (weighting factor).

ANALYSIS RESULTS

In the first trial (Case 1), estimation of C_i is made with α_i=1. From now onward, parameters of $i=L$ represents those for the liquefiable layer and $i=NL$ is for the surface non-liquefiable layer. Estimated combinations of scaling factors (C_i^*) satisfying $\Delta E<0.1$ (residual error of 10%) are plotted in Fig. 2. No unique relationship between C_{NL}^* and C_L^* can be seen: C_L^* are scattered around C_L^*=0.1 and C_{NL}^* lie from 0 to 1 all around. From this figure it can be said that the parameter change for the surface non-liquefiable layer is less sensitive to the pile response than that for the liquefiable layer. For instance, C_{NL}^* can be 0.4 to 0.9 while the range for C_L^* is 0.3 to 0.13 for the test number 2 when we accept the residual error of 10%.

Figure 3 shows the best-fit scaling factors (C^*_{NL} & C^*_L) change with average of the absolute relative displacement between the pile and free-field soil in the layer, |$y_{average}$|. There is a tendency to increase with |$y_{average}$|/D for C^*_{NL}, while C^*_L are again scattered around C_L=0.1. As the strain dependency of the apparent soil stiffness is considered in the form of the hyperbolic p-y curve, C_i should keep constants. However C_i can change when (1) appropriate α_i is not chosen or (2) the hyperbolic function is not suitable for p-y curve modelling. Probably the former is the case for this trial and α_{NL} used may be too large to accommodate the strain dependency of the apparent soil stiffness in the p-y curves.

Based on the above results, in Case 2, estimation of the scaling factor for k_{hr} in the surface non-liquefiable layer, α_{NL}, and the scaling factor for the overall p-y curve in the liquefiable layer, C_L, is made keeping C_{NL} and α_L constants (=1). (As C_{NL}=1, $p_{max} = 3K_p\sigma'_v$ for the surface layer.) Figure 4 plots the best-fit scaling factors (α^*_{NL} & C^*_L) change with the average relative displacement between the pile and free-field soil. When C_{NL} and α_L are kept constants and α^*_{NL} & C^*_L are plotted against |$y_{average}$|/D in semi-log scale, no clear tendency to increase/decrease with |$y_{average}$|/D can be seen for both α^*_{NL} and C^*_L, which means that the strain dependency of the apparent soil stiffness can be accommodated in the form of the hyperbolic p-y curve in Case 2 although amounts of the scatter are still large.

Table 2 summarises the expected parameters for p-y curves in both cases. In estimation of the parameters, (1) distribution of the parameters is assumed to be either logarithmic normal distribution (LND) or normal distribution (ND), and (2) when LND is assumed the data having very small best-fit parameters ($<10^{-4}$) are excluded since the very small relative displacements between the pile and the free-field soil in these particular cases are believed to result in less reliable estimation of the parameter α_{NL}. Set of data obtained gives us confidence limits for the parameters indicating ranges the scaling factors exist, but they do not necessarily assure ability of the parameters to predict the pile response we concern, i.e., the scaling factor parameters within the confidence limits do not always give us good maximum bending moment prediction.

Using the expected scaling factors, calculations are made to evaluate errors in the maximum bending moment (M_{max}) prediction. Errors in M_{max} are plotted against difference between the best-fit and mean scaling factors ($\mu_i^{(mean)}$) normalised by unbiased standard deviation (σ_i) in Figs. 5 and 6. In the figures, plots in upper area represent the maximum bending moment being overestimated and those in left area mean that the soil springs with the mean scaling factors are stiffer/stronger than those with the best-fit parameters. M_{max} can be reasonably predicted as a whole when the best-fit scaling factor is greater than $\mu_i^{(mean)}$, while M_{max} is remarkably overestimated when the best-fit factor is less than $\mu_i^{(mean)}$. However there are some exceptions; M_{max} for test numbers 3, 5 & 7 are underestimated even though the best-fit scaling factor is greater than $\mu_i^{(mean)}$. Possible reasons for this are; (1) difference in deformation mode of the quay wall for No. 3 & 5, and (2) closeness of the pile tip to the quay wall rubble mound for No. 7, as listed in the Table 1.

Additional calculations are made using various combinations of the scaling factors for Case 1 with assumption that logarithm of parameters is normally distributed. Figure 7 plots errors in M_{max} when the values at the upper confidence limits are used for both the non-liquefiable and liquefiable layers ($\mu_{NL}^{(UCL)}$ & $\mu_{L}^{(UCL)}$), et cetera. M_{max} can be reasonably predicted as the mean scaling factors do when the lower confidence limit value is not used for the liquefiable layer. These calculations indicate that (1) the parameter change for the surface non-liquefiable layer is less sensitive to the pile response than that for the liquefiable layer, and (2) conservative prediction of the maximum bending moment in a pile can be reasonably made with the scaling factor greater than the expected mean value determined for the liquefiable layer and those within a range of its confidence limits determined for the surface layer when only the kinematic loadings due to large horizontal movement of soils are considered.

In the calculations made in this study, only the kinematic loadings due to large horizontal movement of soils are considered and conservative prediction of maximum bending moment can be made using stiffer/stronger p-y curves. However, it may be not the case when the superstructure inertial forces are also considered. Expected mean values for the scaling factors would be appropriate for the pile response assessments when both the kinematic and inertial loadings are considered.

Further examinations for piles in laterally spreading soils having various boundary conditions, e.g., piles located behind different types of quay wall, those subjected to horizontal movement of soils having various displacement profiles, etc., may be needed to expand scope of application.

CONCLUSIONS

Variation of p-y curve parameters for piles in laterally spreading liquefied soils is systematically examined by numerically analysing physical model tests undertaken by independent researchers using the beam on non-linear Winkler foundation method with hyperbolic type p-y curves. Estimated best-fit parameters for p-y curve are widely scattered as expected. Key findings obtained are; (1) the parameter change for the surface non-liquefiable layer is less sensitive to the pile response than that for the liquefiable layer, and (2) conservative prediction of the maximum bending moment in a pile can be reasonably made with the scaling factor greater than the expected mean value determined for the liquefiable layer and those within a range of its confidence limits determined for the surface layer when only the kinematic loadings due to large horizontal movement of soils are considered. Further examinations for piles in laterally spreading soils having various boundary conditions may be needed to expand scope of application.

ACKNOWLEDGEMENT

The authors are grateful to Prof S. Yasuda and Dr T. Tanaka of Tokyo Denki University for allowing them to use unpublished physical model tests data.

REFERENCES

Brandenberg, S.J., Singh, P., Boulanger, R.W., and Kutter, B.L. (2001). "Behavior of piles in laterally spreading ground during earthquakes.", *Proc. 6th Caltrans Seismic Research Workshop*, CA, Paper 02-106.

Broms, B.B. (1964). "Lateral resistance of piles in cohesionless soils." *J. Soil Mech. Found., ASCE*, Vol.90, No.SM3, 123-156.

Fujiwara, T., Horikoshi, K., and Sueoka, T. (1997). "Centrifuge model tests on gravity type quay wall." *Proc. Sym. Liquefaction-induced lateral Spreading of Soils*, Tokyo, Japan, 235-240 (in Japanese).

Fujiwara, T., Horikoshi, K., and Sueoka, T. (1998). "Dynamic behavior of gravity type quay wall and surrounding soil during earthquake." *Proc. Centrifuge 98*, Tokyo, Japan, 359-364.

Horikoshi, K., Tateishi, A., and Fujiwara, T. (1998). "Centrifuge modeling of a single pile subjected to liquefaction-induced lateral spreading." *Special Issue of Soils. Found.*, No.2, 193-208.

Japan Road Association (2002). *Specification for Highway Bridges*, Part IV, JRA, Tokyo, Japan.

Japan Society of Civil Engineers (2000). Earthquake Resistant Design Codes in Japan, JSCE, Tokyo, Japan.

Public Works Research Institute (1998a). "Shaking table tests on piles subjected to liquefaction-induced laterally spreading soils." *Internal Report*, PWRI, Tsukuba, Japan (in Japanese).

Public Works Research Institute (1999b). "Dynamic centrifuge model tests on liquefaction remediation for mitigation of damage of piles subjected to liquefaction-induced laterally spreading soils." *Internal Report*, PWRI, Tsukuba, Japan (in Japanese).

Sento, N, Yanagisawa, E., and Fujiki, H. (1998). "1g shaking table tests on different flexural rigidity piles in liquefaction-induced lateral spreading of soils." *Proc. 33rd Japan National Conf. Geotech. Eng.*, Yamaguchi, Japan, 1003-1004 (in Japanese).

Takahashi, A. (2004). "TC29 bender element round robin test results." *Soil Mechanics Section Internal Report*, Imperial College London, UK.

Yasuda, S., Tanaka, T., and Ishii, T. (2004). "Adaptability of pile installation method as a countermeasure against liquefaction-induced flow." *Proc. 15th Southeast Asian Geotech. Conf.*, Bangkok, Thailand, 917-922.

Yasuda, S., and Tanaka, T. (2005). Personal communication.

Table 1: Summary of physical model tests

No.	CA	Soils Surface layer	H_{NL}	Liquefiable layer	H_L	Base layer	Input motion
1	1	Loose fine gravel	0.10	Toyoura sand (Dr=50%)	0.30	Toyoura sand (H=0.25m, Dr=65%)	Sinusoidal waves (0.15G, 10Hz, n=20)
2	1	Loose fine gravel	0.10	Toyoura sand (Dr=50%)	0.30	Toyoura sand (H=0.25m, Dr=65%)	Sinusoidal waves (0.15G, 10Hz, n=20)
3	1	Loose fine gravel	0.10	Nikko silica sand (Dr=50%)	0.40	Nikko silica sand (H=0.27m, Dr=50%)	Sinusoidal waves (0.40G, 3Hz, n=20)
4	1	Loose fine gravel	0.00	Nikko silica sand (Dr=50%)	0.50	Nikko silica sand (H=0.27m, Dr=50%)	Sinusoidal waves (0.40G, 3Hz, n=20)
5	1	Loose fine gravel	0.10	Nikko silica sand (Dr=50%)	0.40	Nikko silica sand (H=0.27m, Dr=50%)	Sinusoidal waves (0.40G, 3Hz, n=20)
6	1	Loose fine gravel	0.10	Nikko silica sand (Dr=50%)	0.40	Nikko silica sand (H=0.27m, Dr=50%)	Sinusoidal waves (0.40G, 8Hz, n=20)
7	1	Toyoura sand (Dr=50%)	0.50	Toyoura sand (Dr=50%)	1.00	Toyoura sand (H=0.3m, Dr=85%)	Sinusoidal waves (0.50G, 5Hz, n=20)
8	50	Toyoura sand (Dr=55%)	0.00	Toyoura sand (Dr=55%)	7.25	--	Sinusoidal waves (0.15G, 1Hz, n=60)
9	50	Toyoura sand (Dr=55%)	0.00	Toyoura sand (Dr=55%)	7.25	--	Sinusoidal waves (0.15G, 1Hz, n=60)
10	50	Toyoura sand (Dr=55%)	0.00	Toyoura sand (Dr=55%)	7.25	--	Sinusoidal waves (0.15G, 1Hz, n=60)
11	50	Toyoura sand (Dr=55%)	3.25	Toyoura sand (Dr=55%)	4.00	--	Sinusoidal waves (0.15G, 1Hz, n=60)
12	50	Toyoura sand (Dr=55%)	3.25	Toyoura sand (Dr=55%)	4.00	--	Sinusoidal waves (0.15G, 1Hz, n=60)
13	50	Toyoura sand (Dr=55%)	1.50	Toyoura sand (Dr=55%)	5.75	--	Sinusoidal waves (0.15G, 1Hz, n=60)
14	50	Toyoura sand (Dr=55%)	1.50	Toyoura sand (Dr=55%)	5.75	--	Port Island-NS @GL-16.4m (0.26G)
15	50	Toyoura sand (Dr=40%)	1.75	Toyoura sand (Dr=40%)	4.25	--	Port Island-NS @GL-16.4m (0.26G)
16	50	Toyoura sand (Dr=40%)	1.75	Toyoura sand (Dr=40%)	4.25	--	Sinusoidal waves (0.19G, 1Hz, n=20)
17	50	Toyoura sand (Dr=40%)	1.75	Toyoura sand (Dr=40%)	4.25	--	Sinusoidal waves (0.10G, 1Hz, n=20)
18*	50	Silica sand No.7 (Dr=60%)	3.00	Silica sand No.7 (Dr=60%)	7.00	Silica sand No.7 (H=10m, Dr=90%)	Sinusoidal waves (0.24G, 1.5Hz, n=24)

CA   : Centrifugal acceleration (G)
H_{NL} : Thickness of surface non-liquefiable layer (m)
H_L : Thickness of liquefiable layer (m)
* Foundation consists of 4 piles rigidly fixed to the pile cap (Pile spacing=2.5D)

Table 2: Summary of physical model tests (continued)

No.	Foundation				Test results			p-y model[†]		p-y model[‡]		Source
	D	EI	βL	s/H	u_q/H	u_g/D	u/D	C^*_{NL}	C^*_L	α^*_{NL}	C^*_L	
1	0.018	1.23E-5	7.33	0.88	--	0.94	1.28	0.00	0.145	0.300	0.100	Sento et al. (1998, Small EI)
2	0.025	9.22E-4	2.60	0.88	--	0.72	0.07	0.65	0.070	0.026	0.070	Sento et al. (1998, Large EI)
3	0.030	2.13E-5	7.98	0.80	0.26**	4.37	4.13	0.00	0.180	0.000	0.185	Yasuda et al. (2004, Case 1)
4	0.030	2.13E-5	9.06	0.80	0.29**	8.27	1.33	--	0.060	--	0.060	Yasuda & Tanaka (2005., Case 2)
5	0.030	1.70E-4	4.85	0.80	0.32**	5.13	1.77	1.00	0.325	0.670	0.325	Yasuda & Tanaka (2005, Case 3)
6	0.030	2.13E-5	7.98	0.80	0.14	1.47	1.83	0.00	0.120	0.000	0.120	Yasuda & Tanaka (2005, Case 4)
7	0.060	5.54E-2	3.01	0.33	0.19	4.58	--	1.00	0.075	0.670	0.085	PWRI (1998a, Case 6)
8	0.325	8.26E+0	5.11	0.28	0.14	3.48	0.83	--	0.090	--	0.090	Horikoshi et al. (1998, Case A)
9	0.325	8.26E+0	5.11	1.31	0.12	2.68	0.89	--	0.095	--	0.095	Horikoshi et al. (1998, Case B)
10	0.325	8.26E+0	5.11	2.00	0.10	2.14	0.38	--	0.075	--	0.075	Horikoshi et al. (1998, Case C)
11	0.325	8.26E+0	4.93	0.28	0.09	1.74	1.46	0.10	0.040	0.003	0.055	Horikoshi et al. (1998, Case E)
12	0.325	8.26E+0	4.93	2.00	0.09	1.20	1.51	0.00	0.245	0.000	0.260	Horikoshi et al. (1998, Case F)
13	0.325	8.26E+0	4.58	0.28	0.10	2.14	0.55	0.15	0.030	0.002	0.025	Horikoshi et al. (1998, Case G)
14	0.325	8.26E+0	4.58	0.28	0.04	0.63	0.48	0.00	0.070	0.002	0.060	Horikoshi et al. (1998, Case H)
15	0.400	5.25E+1	2.41	1.88	0.18	0.99	0.15	0.15	0.180	0.005	0.185	Fujiwara et al. (1998, Case A)
16	0.400	5.25E+1	2.41	1.88	Over-turned	1.90	0.20	0.25	0.135	0.003	0.150	Fujiwara et al. (1997, Case B')
17	0.400	5.25E+1	2.41	1.88	0.23	0.45	0.08	0.15	0.050	0.008	0.070	Fujiwara et al. (1998, Case B)
18*	0.500	3.70E+2	5.08	0.98	0.13	2.00	0.56	0.40	0.005	0.012	0.005	PWRI (1998b, Case 1)

D : Pile diameter (m)
EI : Flexural rigidity of pile (MN.m^2)
βL : Dimensionless length of pile (β=stiffness ratio of soil to pile & L=pile length)
s : Distance of foundation from quay wall (m)
H : $H_{NL} + H_L$ (m)
u_q : Horizontal displacement of top of quay wall (m)
u_g : Horizontal displacement of ground surface at foundation location (m)
u : Horizontal displacement of foundation (m)
C^*_{NL} : Estimated scaling factor for non-liquefiable layer in p-y model
C^*_L : Estimated scaling factor for liquefiable layer in p-y model
α^*_{NL} : Estimated scaling factor for k_{hr} for non-liquefiable layer
** Heal settlement of quay wall is greater than that at the toe.
[†] Best-fit parameters with α_{NL}=1 and α_L=1 (Case 1).
[‡] Best-fit parameters with C_{NL}=1 and α_L=1 (Case 2).

Table 2: Summary of expected parameters for p-y curves

		Case 1				Case 2			
		C_{NL}	α_{NL}	C_L	α_L	C_{NL}	α_{NL}	C_L	α_L
LND	$\mu_i^{(mean)}$	3.0E-1	1	8.2E-2	1	1	1.9E-2	8.3E-2	1
	$\mu_i^{(UCL)}$	5.9E-1	--	1.3E-1	--	--	8.3E-2	1.3E-1	--
	$\mu_i^{(LCL)}$	1.6E-1	--	5.1E-2	--	--	4.2E-3	5.2E-2	--
ND	$\mu_i^{(mean)}$	2.8E-1	1	1.1E-1	1	1	1.2E-1	1.1E-1	1
	$\mu_i^{(UCL)}$	4.8E-1	--	1.5E-1	--	--	2.6E-1	1.5E-1	--
	$\mu_i^{(LCL)}$	7.0E-2	--	7.1E-2	--	--	(-1.9E-2)	7.1E-2	--

LND: Logarithm of parameters is assumed to be normally distributed.
ND: Parameters are assumed to be normally distributed.
$\mu_i^{(mean)}$: Mean value of parameter
$\mu_i^{(UCL)}$: Upper confidence limit with risk degree of 0.05 (confidence coefficient of 0.95)
$\mu_i^{(LCL)}$: Lower confidence limit with risk degree of 0.05 (confidence coefficient of 0.95)

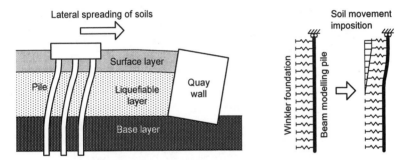

Target pile foundation in laterally spreading soils   Schematic view of analytical model used

Figure 1: Schematic view of problem to be solved

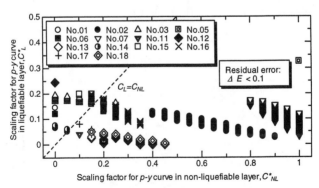

Figure 2: Combinations of scaling factors for p-y curve whose residual error, $\Delta E<0.1$ (Case 1, except Nos. 4, 8, 9 & 10)

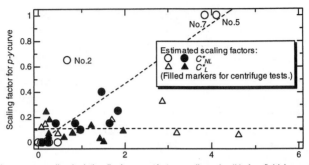

Figure 3: Best-fit *p-y* curve scaling factors (C^*_{NL} & C^*_L) change with average relative displacement between pile and free-field soil (Case 1)

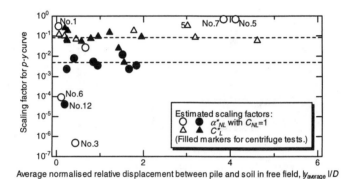

Figure 4: Best-fit *p-y* curve scaling factors (α^*_{NL} & C^*_L) change with average relative displacement between pile and free-field soil (Case 2)

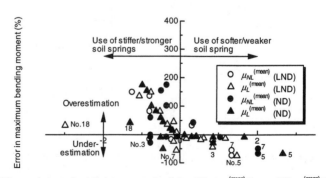

Figure 5: Errors in maximum bending moment with mean scaling factors (Case 1)

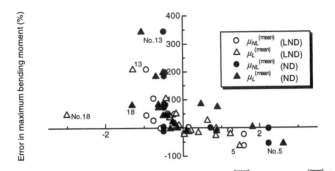

Figure 6: Errors in maximum bending moment with mean scaling factors (Case 2)

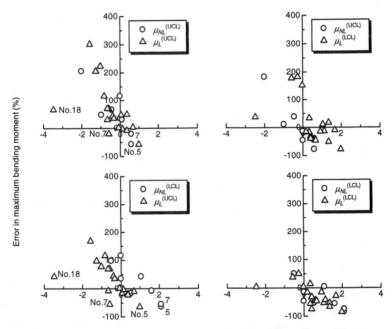

Figure 7: Errors in maximum bending moment with expected scaling factors (Case 1, Logarithm of parameters is assumed to be normally distributed.)

ASSESSMENT OF PILE GROUP RESPONSE TO LATERAL SPREADING BY SINGLE PILE ANALYSIS

Misko Cubrinovski[1], Kenji Ishihara, Member, ASCE[2]

ABSTRACT

In the simplified analysis of piles subjected to lateral spreading, a single pile model is commonly used, and therefore, one of the key issues is how to apply this model to the analysis of pile groups, particularly in the case when individual piles in the group are subjected to different lateral ground displacements. This paper presents a simple concept for analysis of pile groups by using a single pile model. The evaluation of the pile group response by means of the single pile analysis is carried out in two steps, each consisting of several iterative calculations. The objective of the first series of analyses is to estimate the lateral displacement of the pile group at the pile-head level while in the second series of analyses the response of each individual pile is evaluated by considering its particular position within the group.

INTRODUCTION

The most frequently encountered soil profile for piles in liquefied deposits consists of three soil layers where the liquefied layer is sandwiched between a non-liquefied layer at the ground surface and non-liquefied base layer. Liquefaction during strong ground shaking results in almost a complete loss of strength and stiffness in the liquefied soil, and consequent large ground deformation. Particularly large and damaging for piles can be post-liquefaction ground displacements due to lateral spreading. During the spreading, a crust layer of non-liquefied soil is carried along with the underlying spreading soil, and when driven against embedded piles, the crust soil is envisioned to exert large lateral loads on the piles. Thus, the excessive

---

[1] Kiso-Jiban Consultants Co. Ltd., Tokyo, Japan 102-8220.
[2] Tokyo University of Science, Chiba, Japan 278-8510.

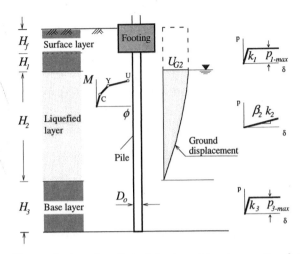

Figure 1: Characterization of nonlinear behavior and input parameters for simplified analysis of piles subjected to lateral spreading

lateral movement of liquefied soils, lateral loads from the surface layer and significant stiffness reduction in the liquefied layer are key features that need to be considered when evaluating the pile response to lateral spreading.

Based on the liquefaction characteristics and kinematic mechanism as above, a three-layer soil model was adopted in a previous study (Cubrinovski and Ishihara, 2004) for a simplified analysis of piles. In the adopted pseudo-static approach, the spreading is represented by a horizontal displacement of the liquefied soil and particular attention is given to the loads arising from the non-liquefied layer at the ground surface and to the modeling of the interface between the liquefied layer and non-liquefied base layer. The analysis permits to evaluate the inelastic response of piles, yet it is based on a closed-form solution and requires only few conventional engineering parameters as input, as illustrated in Fig. 1. Key parameters influencing the pile response have been identified to be the magnitude of lateral ground displacement U_{G2}, ultimate pressure form the crust layer p_{1-max} and stiffness reduction in the liquefied soil β_2 (Cubrinovski and Ishihara, 2004; Cubrinovski et al., 2004).

As commonly adopted in the simplified analyses of piles, the above method is based on a single pile model, and hence one of the key issues is how to apply this method to the analysis of pile foundations where a large number of piles are clustered in a group and rigidly connected at the pile head. The interaction among the piles is of particular importance for pile foundations in waterfront areas where individual piles within the group are subjected to different lateral ground displacements. This paper focuses exclusively on this issue and presents a simple concept for analysis of pile groups by using a single pile model.

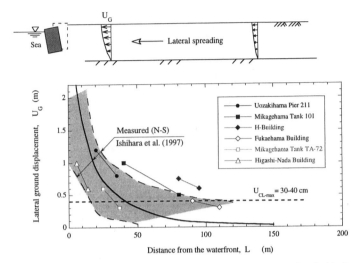

Figure 2: Permanent lateral ground displacements due to spreading behind quay
walls in the 1995 Kobe earthquake

LATERAL SPREADING BEHIND QUAY WALLS

Distribution of Lateral Ground Displacements

During the 1995 Kobe earthquake, lateral spreading of liquefied soils occurred in
the waterfront areas of many artificial islands. The quay walls moved several meters
towards the sea and lateral spreading of the liquefied backfills extended inland as far
as 200 m from the revetment line. Ishihara et al. (1997) investigated the features of
movements of the quay walls and ground distortion in the backfills by ground
surveying measurements along alignments in the direction perpendicular to the
revetment line. They summarized the measured displacements in plots depicting the
permanent ground displacement as a function of the distance inland from the
waterfront, as shown in Fig. 2 where the shaded area shows the range of measured
displacements along N-S sections of Port Island and the solid line is an
approximation for the average displacement. The lateral ground displacement is
largest at the quay walls and decreases rapidly with the distance from the waterfront.
This is typical distribution of permanent ground displacements due to lateral
spreading in waterfront area observed in many earthquakes and adopted in several
empirical models (e.g. Bartlett and Youd, 1995; Tokimatsu and Asaka, 1998).

A large number of buildings, storage tanks and bridge piers on pile foundations
were located within the lateral spreading zone as depicted in Fig. 2 where the
location of the end piles of a given foundation are indicated, that is the pile which is
the closest to the quay wall and the pile on the inland side which is the farthest away

from the waterfront. It is apparent that, depending on the location within the pile group, different piles of a given foundation were subjected to significantly different lateral ground displacements.

Effects on Pile Groups

In order to illustrate this feature of lateral spreading displacements and its effects on the behavior of pile groups, a large pile foundation consisting of n-piles located within the lateral spreading zone is considered in the following, as shown schematically in Fig. 3a. The *Pile n* is closest to the quay wall while *Pile 1* is on the inland side, furthest from the revetment line. Assuming typical distribution of spreading displacements in which the magnitude of lateral ground displacement U_G decreases with the distance from the waterfront, the lateral ground displacement at a distance corresponding to the location of Pile 1 will be $U_{G(1)}$ while the corresponding displacement at Pile n will be $U_{G(n)}$ where $U_{G(1)} < U_{G(2)} < ... < U_{G(n)}$. Note that here U_G denotes a free field displacement at the ground surface which is assumed to be identical to the displacement at the top of the liquefied layer. In the simplified methods for analysis of piles based on the displacement approach, this free field ground displacement is commonly used as an input, and hence, each pile of the foundation in question will be subjected to a different lateral ground displacement, as illustrated in Fig. 3b. Thus, when considering the piles individually, each of them will tend to move a different amount, in proportion to the lateral ground displacement or applied soil pressure. However, since the piles are rigidly connected at the head through a foundation mat or a system of pile caps and underground beams, the whole foundation will move laterally as a unit and therefore all of the piles will share nearly identical horizontal displacements at the pile head. This cross-interaction among the piles through the pile-foundation-pile system affects both the overall movement of the foundation as well as the bending deformation of the

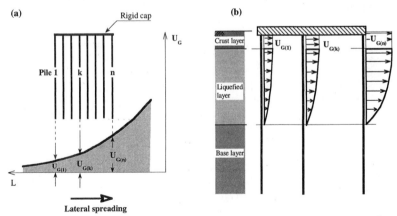

Figure 3: Piles in a group subjected to lateral ground displacements due to spreading

individual piles. These effects can be particularly large in cases when the ground displacement changes significantly with the distance from the waterfront which is a feature commonly present in the close proximity to the quay walls and for foundations having a relatively large dimension in the direction of spreading. Tokimatsu and Asaka (1998) and Tokimatsu and Suzuki (2002) investigated the cross-interaction of piles in a group as above and demonstrated its effects in the performance of pile foundations during the 1995 Kobe earthquake. They proposed an analytical method using a frame model for the pile group that permits to evaluate the pile response by taking into account the interaction effects as above.

In the present study, we investigated the possibility to consider these effects and hence evaluate the response of pile groups using an analysis with a single pile model, as described in the following.

SIMPLIFIED ANALYSIS OF PILE GROUP WITH A SINGLE PILE MODEL

As mentioned above, piles in a group are rigidly connected at the pile head, and therefore, when subjected to lateral loads, the top part of the pile foundation will move as a unit and all piles will share nearly identical horizontal displacements at the pile head. On the other hand, each of the piles will be subjected to a different lateral load from the surrounding soils depending upon its particular location within the group and spatial distribution of lateral ground displacements. Consequently, both the interaction force at the pile head and the lateral soil pressure along the length of the pile will be different for each pile thus leading to a development of distinct deformation and stresses along the length of individual piles in the group. This response feature in which the piles share identical displacements at the pile head but have different deformations throughout the depth is considered to be the outlining feature of the deformational behavior of pile groups subjected to lateral spreading. Thus, the principal goal of the single pile analysis is to capture this behavior of piles in a group.

In accordance with the above-mentioned interaction mechanism, the evaluation of the pile group response by means of a single pile analysis is carried out in two steps (series of analyses), each consisting of several iterative calculations. The objective of the first series of analyses (Step 1) is to estimate the lateral displacement of the pile group at the pile-head level. In the second series of analyses (Step 2), the bending moments, shear stresses and lateral loads are computed for each pile individually by considering its particular position within the group. In what follows, the proposed analysis procedure is illustrated through an example calculation of a simplified pile group configuration.

The pile foundation in question consists of 5 piles which are rigidly connected at the pile head, as shown schematically in Fig. 4. The piles are embedded in a three-layer deposit with a non-liquefied crust layer overlying the liquefied layer and a non-liquefied base layer. Bilinear and equivalent linear p-δ relationships are adopted for the non-liquefied layers and the liquefied layer, respectively, while a trilinear

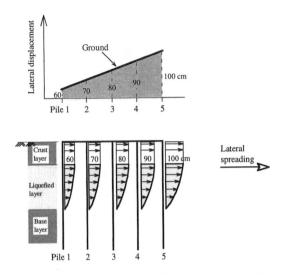

Figure 4: Pile foundation considered in the example calculation

moment-curvature relationship is used for the pile. Thus, the stress-deformation relationships of the soil and the pile are identical to those of the simplified model shown in Fig. 1. It is assumed that the piles are subjected to a lateral ground displacement which linearly increases in the direction of spreading from 60 cm, at the location of *Pile 1*, to 1.0 m at the location of *Pile 5*, as depicted in Fig. 4. Note that these are the lateral ground displacements at the top of the liquefied layer which are assumed to be identical with the corresponding displacements at the ground surface.

Lateral Displacement of the Pile Group

The calculation of the lateral displacement of the pile group by means of the single pile analysis is summarized in Fig. 5. The procedure essentially consists of an analysis with the single pile model and several simple calculations used for identifying the appropriate ground displacement and ultimate pressure from the crust layer to be used as input in this analysis. Note that this single pile analysis is aimed at evaluating the pile group displacement at the pile head, and hence, it is basically an analysis in which the average lateral loads per pile are applied. In other words, the total lateral load obtained by the superimposition of the lateral loads on all piles in the group is divided by the total number of piles, and thus defined lateral load is applied to a single pile. The procedure outlined in Fig. 5 shows how to evaluate the average loads from the liquefied layer and crust layer which will eventually lead to a lateral displacement of the single pile which is identical to that of the pile group.

As shown in Fig. 5, first the average ground displacement $\overline{U}_G$ is computed by

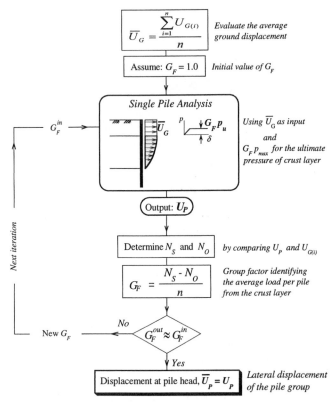

Figure 5: Calculation of lateral displacement of pile group with a single pile model
(Step 1)

$$\overline{U}_G = \frac{\sum_{i=1}^{n} U_{G(i)}}{n} \tag{1}$$

where n is the total number of piles ($n = 5$ in this case) and $U_{G(i)}$ is the lateral ground displacement corresponding to the location of each pile. Using this average ground displacement as an input, an analysis is then conducted with the single pile model in which the ultimate pressure from the crust layer is defined by the product $G_F \cdot p_u$ where $G_F = 1.0$ is initially assumed.

Suppose that the lateral displacement at the pile head U_P computed in this analysis is larger than the ground displacement $U_{G(1)}$, but smaller than $U_{G(2)}$, as indicated in Fig. 6a. Assuming that the computed pile displacement U_P is actually the horizontal displacement of the pile group, then, the displacement of *Pile 1* will be greater than

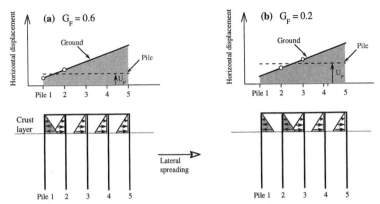

Figure 6: Illustration of lateral loads from the crust layer on piles in a group:
(a) $U_{G(1)} < U_P < U_{G(2)}$; (b) $U_{G(2)} < U_P < U_{G(3)}$

the displacement of the surrounding soil, and therefore, the lateral pressure from the crust soil on *Pile 1* will be in the direction opposite to the lateral spreading. On the other hand, the pressure from the crust soil on *Piles 2 to 5* will be in the spreading direction, as depicted in Fig. 6a. Assuming that the ultimate lateral pressure from the crust soil has been mobilized on all piles, then the average load from the crust layer per pile will be proportional to a factor G_F given by

$$G_F = \frac{N_s - N_o}{n} \qquad (2)$$

where N_S is the number of piles that are subjected to a lateral pressure from the crust soil in the direction of spreading ($U_P < U_{G(k)}$, for these piles), while N_O is the number of piles that are pushed by the crust soil in the direction opposite to that of the spreading ($U_P > U_{G(j)}$, for these piles). For the particular case illustrated in Fig. 6a, $G_F = (4-1)/5 = 0.6$, and hence, the average load from the crust layer per pile will be 60% of the maximum load, in the direction of lateral spreading. Another possible outcome of the calculation is depicted in Fig. 6b, in which case $U_{G(2)} < U_P < U_{G(3)}$, and hence, $G_F = (3-2)/5 = 0.2$. It is important to mention that Eq. (2) is applicable only over the range of pile displacements for which the ultimate lateral pressure has been mobilized for all piles while in the case when the lateral pressure from the crust layer is in the elastic range, for a given pile, then the group factor is given by

$$G_F = \frac{N_s - N_o}{n} - \frac{1}{n} \frac{U_P - (U_{G(j)} \mp \delta_y)}{\delta_y} \qquad (3)$$

This feature is illustrated in Fig. 7 where the change in G_F for the considered group of five piles is shown as a function of the lateral displacement of the pile group.

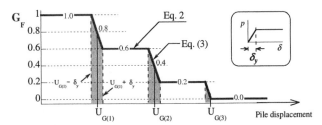

Figure 7: Variation of G_F with the lateral displacement of pile group

Thus, by comparing the pile displacement U_P computed in the single pile analysis with the assumed ground displacement over the area of the pile foundation, one can easily estimate the appropriate group factor G_F and hence evaluate the size of the average load from the crust layer per pile. As indicated in Fig. 5, the value of G_F is adjusted through iterative calculations until eventually the G_F resulting from the analysis is nearly equal to the value of G_F used as an input in the analysis ($G_F^{out} \approx G_F^{in}$). It is important to note that the procedure described above essentially specifies the average lateral loads per pile from the liquefied soil, as defined by the average ground displacement $\overline{U}_G$, and from the crust layer, as defined by the group factor G_F. The displacement at the pile head U_P computed in the last analysis is practically identical with the corresponding lateral displacement of the pile group, $\overline{U}_p$.

Response of Individual Piles in the Group

Once the horizontal displacement of the pile group $\overline{U}_p$ has been identified, the response of each individual pile can be evaluated by an additional series of analyses with the single pile model, as shown schematically in Fig. 8. Here, the calculation is illustrated on the evaluation of the response of *Pile k* in the group where the free field lateral ground displacement at the top of the liquefied layer is $U_{G(k)}$ respectively.

In this calculation procedure, first an analysis is conducted in which the actual ground displacement corresponding to the pile in question, in this case $U_{G(k)}$, is applied as an input. Note that the ultimate lateral pressure of the crust soil p_u corresponding to a single pile is used in this analysis without considering the pile group effects or G_F. This is essentially an analysis in which the actual loads are applied to the pile in question, but by considering it as an isolated single pile and thus ignoring the presence and effects of the other piles in the group.

In the general case, the pile-head displacement U_P computed in the first analysis will be different from the displacement of the pile group $\overline{U}_p$ because the former does not account for the interaction force at the pile head that is transferred to the pile through the upper foundation (cap). However, the difference between these two displacements will reveal the direction of the interaction force, which is denoted by F

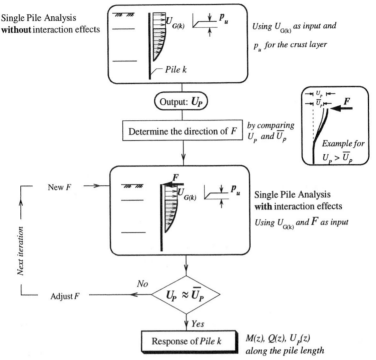

Figure 8: Calculation of the response of individual piles in the group (*Pile k*) with the single pile model (Step 2)

in the following. Namely, if U_P is smaller than $\overline{U}_p$ then F will be in the direction of spreading while, conversely, F will act in the direction opposite to the lateral spreading when $U_P > \overline{U}_p$. The latter case is schematically illustrated in Fig. 8. Once the direction of the interaction force has been identified, the correct magnitude of this force can be easily evaluated in a trial and error procedure in which the value of F is systematically modified and adjusted until eventually the computed displacement at the pile head is identical to the displacement of the pile group $\overline{U}_p$, which serves as a target response in these trial and error calculations. One can interpret F as the lateral force required to be applied at the pile head so that the displacement of *Pile k* is equal to $\overline{U}_p$, and hence, it is equivalent to the interaction force that brings the pile to move together with the group. The response computed in the last analysis, in which the computed displacement at the pile head is equal to $\overline{U}_p$, provides the specific deformation and stresses along the length of *Pile k*. In this way, the response of any pile in the group can be evaluated by means of the single pile analysis.

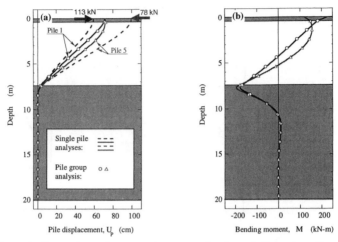

Figure 9: Comparison of computed responses for the end piles using pile group analysis and single pile analysis: (a) Pile displacements; (b) Bending moments

RESPONSE OF END PILES

The simplified analysis of pile groups by way of the single pile method described above provides exactly the same response for the piles in a group as the one obtained from a more detailed analysis using a pile group model. In order to illustrate the accuracy of the calculation and some important features of the response of piles in a group, results of pile group analysis and the proposed single pile analysis of the pile foundation shown in Fig. 4 are briefly discussed below. Here, only the responses of the end piles (*Pile 1* and *Pile 5*) are discussed since these are the two extreme responses within the pile group. Both the pile group analysis and the single pile analysis were conducted using a beam-spring FE models with a series of beam members and horizontal springs representing the piles and soil respectively. In the case of the pile group analysis, the five piles were rigidly connected at the top and were subjected to different lateral ground displacements ranging from 60 cm to 100 cm, as shown in Fig. 4. In the case of the single pile analysis, a single pile model with a constrained rotation at the top was used, and the procedure described above was followed. It is to be noted that the above FE analyses served to rigorously examine the accuracy of the proposed single pile method of analysis, and therefore, an identical methodology was adopted for the pile group and single pile analyses. The simplified method of analysis based on a closed form solution (Cubrinovski and Ishihara, 2004) is actually intended to be used, because of its advantages in conducting parametric evaluation and addressing the uncertainties involved.

Computed lateral displacements and bending moments along the length of *Pile 1* and *Pile 5* are shown in Fig. 9. Here, the symbols indicate the response computed using the pile group model while the solid lines show the response obtained with the proposed single pile analysis. The broken lines in Fig. 9 indicate the computed pile displacements in the analysis without considering the interaction force (illustrated at the top of Fig. 8), while the arrows indicate the interaction forces that are eventually applied to the piles in order to achieve the group pile displacement $\overline{U}_p$, as previously described and illustrated in Fig. 8. Apparently, the forces acting at the pile head are in the opposite direction for *Piles 1* and *5* thus leading to different bending moments especially in the upper part of the piles. It is interesting to notice that the peak bending moment for *Pile 5* occurred within the liquefied layer which is consistent with some of the damage to piles observed in the 1995 Kobe earthquake and is a response feature that can not be explained by a conventional single pile analysis.

CONCLUDING REMARKS

During lateral spreading of liquefied soils in the waterfront area, pile foundations are subjected to spatially varying ground distortion where individual piles within the group are subjected to different lateral ground displacements. The combination of different lateral loads along the pile length and interaction forces at the pile head results in distinct deformational features of each individual pile depending upon its particular position within the group. In this paper, a simplified method for analysis of pile groups and effects as above by a single pile model has been presented. The evaluation of the pile group response by the single pile analysis is conducted in two steps. In the first step, the lateral displacement of the pile group is evaluated by applying the average lateral loads from the liquefied layer and crust layer per pile while in the second step, the response of individual piles is computed by considering the particular lateral loads along the pile length and inertial force at the pile head depending upon the position of the pile within the group. The proposed single pile analysis permits to evaluate the response of pile groups and their specific deformational features that can not be otherwise explained by a conventional single pile analysis.

REFERENCES

Bartlett, S.F. and Youd, T.L. (1995). "Empirical prediction of liquefaction-induced lateral spread." *ASCE J. of Geotech. Engrg.*, Vol. 121, No. 4, 316-329.

Cubrinovski, M., and Ishihara, K. (2004). "Simplified method for analysis of piles undergoing lateral spreading in liquefied soils." *Soils and Foundations*, Vol. 44, No. 5, 119-133.

Cubrinovski, M., Kokusho, T. and Ishihara, K. (2004). "Interpretation from large-scale shake table tests on piles subjected to spreading of liquefied soils." *Proc. 11th Int. Conf. Soil Dynamics and Earthq. Engrg. / 3rd Int. Conf. Earthq. Geotech. Engrg.*, Berkeley, USA, Vol. 2, 463-470.

Ishihara, K., Yoshida, K. and Kato, M. (1997). "Characteristics of lateral spreading in liquefied deposits during the 1995 Hanshin-Awaji earthquake." *Journal of Earthquake Engineering*, Vol. 1, No. 1, 23-55.

Tokimatsu, K. and Asaka, Y. (1998). "Effects of liquefaction-induced ground displacements on pile performances in the 1995 Hyogoken-Nambu earthquake." *Special Issue of Soils and Foundations*, 163-177.

Tokimatsu, K. and Suzuki, H. (2002). "Performance of cast-in-place concrete piles subjected to lateral spread during the 1995 Kobe earthquake." *Workshop on Performance of Ground and Pile Foundations during Soil Liquefaction and Lateral Ground Deformation*, Tokyo, Japan (presentation materials).

MODELLING OF LIQUEFACTION-INDUCED INSTABILITY IN PILE GROUPS

Jonathan A. Knappett[1] and S. P. Gopal Madabhushi[2]

ABSTRACT

Centrifuge testing has been undertaken to investigate instability failure of pile groups during seismic liquefaction, with specific reference to the 'top-down' propagation of liquefaction during the earthquake and to account for initial imperfections in pile geometry. The results of these tests were used to validate numerical models within the finite element program ABAQUS, based on the popular p-y analysis method. Pseudostatic classical and post-buckling analyses were conducted to examine the collapse behaviour of the pile groups and were found to give reasonable predictions of collapse load and conservative predictions of the associated deflection conditions. This numerical model was compared to currently published methods which were found to over-predict collapse loads. The resulting insights into the collapse of axially loaded pile groups revealed that the failure load is strongly dependent on both the depth of liquefaction propagation and initial imperfections, which reduce the collapse load.

INTRODUCTION

Previous work at the University of Cambridge by Bhattacharya et al. (2004) has demonstrated that end-bearing piled foundations can be vulnerable to instability failure during seismic liquefaction based on the results of centrifuge tests. This work considered single piles and suggested that instability occurs at full liquefaction if the

---

[1] Research student, Schofield Centre, University of Cambridge, UK
[2] Senior Lecturer, Schofield Centre, University of Cambridge, UK

applied axial load is greater than the elastic critical (Euler) load in this unsupported condition. The aim of the work presented herein is to extend this work to end-bearing pile groups with specific reference to the 'top-down' propagation of liquefaction during the earthquake and to account for initial imperfections in pile geometry which commonly manifest themselves as an offset deflection between the tip and top of the pile (e.g. due to pile wander during installation). Previous numerical modelling of pile buckling in non-liquefiable soils by Hoadley (1974) has suggested that even small imperfections can have a dramatic reducing effect on the collapse load.

The work described in this paper considers axial load effects on piles alone; lateral loading effects due to soil flow are not considered. Other researchers such as Boulanger et al. (2003) and Haigh & Madabhushi (2005) consider the effects of lateral loading on piles exclusively.

CENTRIFUGE MODELLING

Methodology

A series of centrifuge tests were conducted at 80-g to determine the collapse behaviour of axially loaded pile groups using the 10m diameter Turner Beam Centrifuge at the Schofield Centre, Cambridge University. The first pile group configuration tested (test JK-03) is shown in Figure 1 and consisted of a 2×2 square group of model piles constructed from an extruded aluminium-alloy tube section, connected by a rigid aluminium-alloy pile cap that was free to translate, but with rotational fixity at the pile-cap interface. The tips of the piles were fixed into a rigid aluminium-alloy base plate, with both translational and rotational fixity as a model of rock-socketed piles. Inertial movements of the pile cap and structure in the direction of shaking were removed by using frames with linear PTFE bearings as shown in Figure 1. This confined pile head displacements to be monotonic in the y-direction instead of cyclic in the x-direction, thereby simplifying the problem by isolating the phenomenon of instability from that of soil-structure interaction.

The axial load was applied to the piles by attaching removable brass weights to the pile cap. In the initial tests, two identical pile groups were tested per flight, such that results for two different axial load configurations could be obtained in a single earthquake. The geometric parameters of the pile groups are given in Table 1 at prototype scale. Comparison with a typical 0.5m diameter steel pipe pile after Fleming et al. (1992) show that bending stiffness and outer diameter are well matched, with these parameters governing instability and lateral soil-pile interaction respectively.

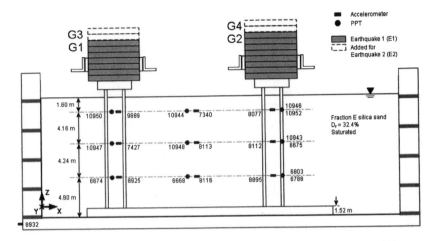

Figure 1: Schematic of layout and instrumentation, centrifuge test JK-03

TABLE 1. Prototype characteristics of model pile groups

Property	Groups G1 – G4 (JK-03)	Groups G5, G6 (JK-04 and -05)	Typical steel pipe pile
Outside diameter (D_0)	0.496 m	0.496 m	0.508
Thickness (t)	0.132 m	0.132 m	0.019
Length (L_p)	14.4 m	8.0m	—
Radius of gyration (r_g)	0.136 m	0.136 m	0.173 m
Youngs Modulus (E)	70 GPa	70 GPa	210 GPa
Yield stress (σ_y)	230 MPa	230 MPa	275 – 355 MPa
Fully-plastic moment (M_p)	4204 kNm	4204 kNm	1250 – 1614 kNm
Cross-sectional area (A)	0.150 m^2	0.150 m^2	0.029 m^2
Flexural stiffness (EI)	189.24 MNm^2	189.24 MNm^2	183.49 MNm^2
Superstructure mass $\{$	2275, 2601, 2926, 3251 Mg	2926, 3901 Mg	—
Axial load per pile (P_p) $\{$	5.58, 6.38, 7.18, 7.97 MN	14.35, 19.14 MN	2.15 – 3.02 MN

The piles and the base plate (without the pile caps) were attached to the bottom of an equivalent shear beam (ESB) container and a loose layer of Fraction E silica sand was pluviated around the piles at a relative density (D_r) of $\approx$ 35%. The model was then saturated under vacuum pressure with a methyl-cellulose/water solution at a viscosity of 80cS in order to provide correct scaling of pore fluid flow – see Schofield (1981). An initial set of brass weights were attached to the pile groups and the model was subjected to a single sinusoidal tone-burst earthquake at 80-g using the stored angular momentum (SAM) earthquake actuator. Further details concerning the design and capacity of this actuator can be found in Madabhushi et al. (1998). The ground motion used had a peak acceleration $|a| \approx 0.1\%g$ at a prototype frequency of 0.63Hz. The initial values of axial load (groups G1 and G2) were insufficient to cause failure, with pile displacements remaining elastic. Additional axial load was therefore added to both pile groups (groups G3 and G4) and the test repeated. Axial load magnitudes are given in Table 1.

Initial imperfections in the piles were incorporated by allowing the pile caps and superstructures to be free to displace during swing up. Swing up occurs when the centrifugal acceleration is approximately 10-g. At this stage, the 1-g component acting in the y-direction displaces the pile cap. As the g-level is increased, the soil stiffness increases and 'locks' the deflected piles in place.

Subsequent tests, JK-04 and JK-05 were performed using similar ground conditions, but with shorter piles, as shown in Table 1. As a result of the reduced length, the axial failure load was expected to be much higher. Due to practical considerations, a 2×1 configuration was used to reduce the magnitude of the load required, with the inertial frames preventing the group from failing in the 'weak' x-direction (in the plane of shaking). The models were subjected to the same earthquake conditions as test JK-03.

Test Observations and Discussion

Full liquefaction occurred during all of the earthquakes, which was confirmed by excess pore pressure measurements in the free field. Deflections were monotonic in the direction of the initial imperfection (as constrained by the inertial frames), though dilation generated by this motion and re-liquefaction in subsequent cycles was observed in the deflection and pore pressure measurements, particularly in the upper levels in the soil layer where the deflection is expected to be greater.

Of the six pile groups tested, group G4 suffered large movements during the earthquake followed by collapse post-earthquake, while group G6 collapsed almost immediately after shaking commenced. Failure modes observed following post-test excavation are shown in Figure 2.

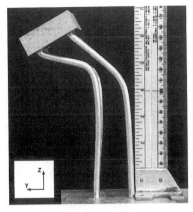

(a) Group G4, JK-03   (b) Group G6, JK-05

Figure 2: Post-test observations of failure modes

NUMERICAL MODELLING

Coupled Load-Deflection Analysis Using a Winkler Soil Model in ABAQUS

The finite element code ABAQUS was used to analyse the foundation numerically, with the soil modelled by non-linear elastic p-y springs at discrete spacings along the piles. ABAQUS has the capacity to solve the equilibrium equations using large deflection theory and perform coupled load-deflection analyses using the modified Riks method as detailed by Crisfield (1981). This allows the unstable (negative stiffness) regions of the load-deflection curve for a structural system to be computed, thereby allowing the full post-buckling behaviour to be determined. This permits the study of bifurcation problems such as the pile group collapse phenomenon observed in the centrifuge tests.

Non-linear Elastic Soil-pile Interaction Behaviour

The pile group was modelled as an elastic-perfectly plastic material with the same properties as in Table 1 and a fully rigid construction was used to model the pile cap and superstructure. The pile-soil interaction behaviour was modelled by considering the load-deflection relationship for a laterally loaded rigid disc in a semi-infinite elastic medium under plane strain conditions presented by Baguelin et al. (1977). This has been successfully used in dynamic p-y analyses for single piles and pile groups without axial load by Maheetharan (1990), and validated by centrifuge testing with a similar silica-based sand at Cambridge University. The resulting load-deflection relationship for the discrete springs is given by:

$$\frac{p}{y} = \frac{16\pi(1-v)GL_s}{\left[(3-4v)\ln\left(\frac{2R}{D_0}\right)^2 - \frac{2}{(3-4v)}\right]}$$ (1)

where G is the secant shear modulus at the given value of y, p represents the spring reaction force, L_s is the length of the section of pile acted upon by the discrete spring and R is the radius of the outer rigid boundary of the soil layer, here taken as R = 17.5D_0 representing the smallest distance to the walls of the ESB container. The non-linear variation of G with shear strain (γ) is assumed to be given by Hardin & Drnevich (1972). In computing the shear modulus variation in this manner, the maximum shear stress (τ_{max}) is taken as the average of the varying maximum shear stress within 2.5D_0 of the pile centreline after Barton (1982):

$$\tau_{max} = \frac{1}{2D_0} \int_{r_0}^{5r_0} \frac{(\sigma_r' - \sigma_\theta')}{2} \, dr$$ (2)

where σ_r' and σ_θ' are the radial and tangential stress distributions in the annular soil element.

Modelling of Liquefaction Propagation in the p-y-u Framework

The key features of the liquefaction process that need to be modelled to replicate the failures observed in Figure 2 are the loss in shear stiffness and the progression of liquefaction as a 'top-down' phenomenon. As the shear modulus is dependant on the level of effective stress, the reduction in stiffness can be accounted for by replacing G_{max} and τ_{max} in the previous equations by $G_{max,u}$ and $\tau_{max,u}$ given by:

$$G_{max,u} = G_{max}(1-r_u)^{\frac{1}{2}}$$ (3)

$$\tau_{max,u} = \tau_{max}(1-0.99r_u)$$ (4)

where r_u is the excess pore pressure ratio. Due to the coupling of load and deflection in the modified Riks method, incrementation for this form of analysis is based on a proportional load magnitude scale factor, as opposed to the more usual time incrementation. As a result, dynamic effects such as liquefaction propagation must be modelled pseudostatically. The value of r_u in the above equations was computed for each soil spring based on a synthetic excess pore pressure profile representing the amount of liquefaction at a particular time instance. Following the method of Brennan (2004), the profiles were assumed to be given by a Fourier series:

$$u(z) = \sum_{n=odd} A_n \sin\left(\frac{n\pi z}{2H}\right) \tag{5}$$

with $n = 1 - 5$ and z measured from the ground surface. This function for $u(z)$ satisfies the required boundary conditions $u(0) = 0$ and $u'(L_p) = 0$. Pseudostatic profiles were computed varying from $r_u = 0$ everywhere to $r_u = 1$ everywhere and are shown in Figure 3 for the case of groups G1 – G4. Corresponding p-y relations for the soil springs in each profile were then computed and coupled load-deflection analyses undertaken for each set of springs, with the load-deflection curve being obtained for each.

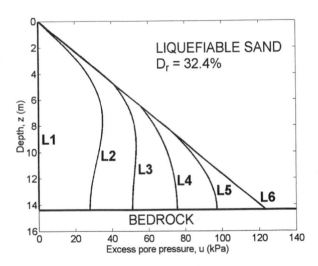

Figure 3: Synthetic excess pore pressure profiles (14.4m long piles)

INSIGHTS INTO PILE GROUP INSTABILITY

Effect of Initial Imperfections and Liquefaction Propagation

Based on currently available methods (Bhattacharya et al., 2004) instability will be avoided if critical loads at full liquefaction (i.e. no soil support along L_p) are higher than the applied loads. The critical load for the pile groups in Table 1 is given as $4P_{cr}$, where P_{cr} is given by:

$$P_{cr} = \frac{\pi^2 EI}{(\beta L_p)^2} \tag{6}$$

For the pile group conditions tested (sway permitted), the effective length factor β = 1.0. Corresponding values of P_{cr} per pile are 9.01MN and 29.18MN for the 14.4 and 8.0m long piles respectively. Table 2 shows ratios of axial to critical loads for the pile group configurations tested. Given these conditions, groups G4 and G6 should not have failed as the Euler loads (based on the full liquefaction depth) are greater than the applied loads. However, it has been shown in the previous section that these pile groups did indeed fail. This therefore demonstrates the importance of initial imperfections.

TABLE 2. Load levels as a function of P_{cr}

Group ID	G1	G2	G3	G4	G5	G6
P_p/P_{cr}	0.62	0.71	0.80	0.88	0.49	0.66

To extend the above elastic method to account for the reduction in collapse load during liquefaction propagation, Eulerian critical loads were also determined for the numerical models for each amount of liquefaction (L1 – L6) by classical eigenvalue prediction in ABAQUS and first mode values are plotted in Figure 4 for the 14.4m long configuration in which the value of r_u at the base of the sand layer ($r_{u,base}$) is used to describe the amount of liquefaction propagation. .

In order to incorporate initial imperfections, elasto-plastic post-buckling analyses were computed for the imperfect pile groups using the modified Riks method in ABAQUS. The model was firstly displaced to the initial deflections observed in the centrifuge tests by applying an equivalent co-linear horizontal force at the centre-of-mass of the structure, with the soil springs removed. This represents the initial deflections occurring at the swing-up stage with the soil at a low effective stress and stiffness level. The soil springs were then replaced in a strain free condition to model the stage between 10-g and 80-g where the soil stiffens up to an assumed G_{max} condition around the deflected piles. The modified Riks method was then implemented by applying the axial load, which was proportionally scaled throughout the analysis (i.e. treated as a variable), and solving additionally and simultaneously for deflection. The load-deflection curves obtained from the post-buckling analysis showed an initially stable region followed by an unstable region after a point of maximum load (bifurcation point) had been reached. Peak loads for each profile of excess pore pressure are shown in Figure 4 for Group G4.

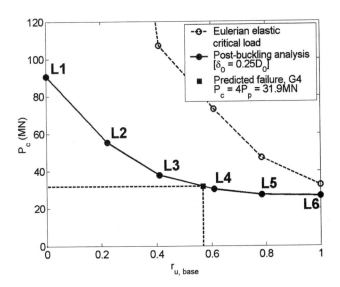

Figure 4: Reduction in collapse load due to liquefaction propagation and imperfections
(Group G4)

The large difference between the Eulerian and post-buckling solutions shown in Figure 4 demonstrates the importance of pile group imperfections in determining the collapse load in this mode. The loading applied to group G4 suggests that the pile group would fail with a liquefaction profile between L3 and L4 as indicated in Figure 4. The load-deflection curves for the analysis of group G4 are shown in Figure 5 along with points detailing the formation of plastic hinges at locations indicated by the diagram in the top right hand corner of the figure. It can be seen that as liquefaction progresses, the equilibrium path progressively softens from L1 towards L6 (in reality this change would be continuous). The equilibrium points at which the plastic hinges form in the piles are also shown in the figure, revealing that hinges form immediately below the pile cap prior to the peak load being reached in all cases. The centrifuge tests are effectively load controlled and the loading path for group G4 is shown on Figure 5, suggesting three regions of behaviour:

- O-A: The pile group deflects in a stable fashion.
- A-B: The upper set of plastic hinges form at point A, changing the fixity condition at the pile heads and causing a dramatic loss of structural stiffness.
- B-F: The change in fixity causes the peak load point to be reached soon after A and the pile group collapses due to bifurcation instability at B.

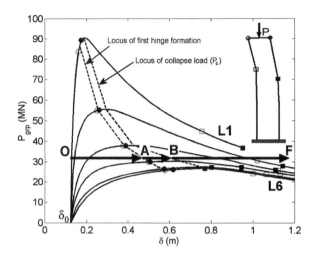

Figure 5: Load-deflection curves, Group G4 ($\delta_0 = 0.25D_0$)

Validation of Numerical Predictions with Centrifuge Test Results

ABAQUS results for all of the pile group configurations tested are summarized in Tables 3 and 4 along with salient values of load and maximum achieved deflection (δ_{max}) observed in the centrifuge tests. δ_A and δ_B represent the deflections at formation of the upper hinges and at collapse respectively.

The ABAQUS model correctly predicts that groups G1, G2 and G5 should not have failed, as the loadings applied to these groups in the centrifuge tests were lower than the lowest value of P_c (i.e. at full liquefaction, L6). The numerical results also suggest that group G3 should have been on the margin of collapse. However, in order for collapse to occur the displacement condition δ_B must also be exceeded and it is clear from Table 3 that $\delta_{max} \ll \delta_B$ (and also δ_A) as a result of dilation-induced resistance in the soil. This effect has also been observed by Haigh & Madabhushi (2005) who measured excess pore pressures in close proximity to piles during lateral shaking.

TABLE 3. Comparison of ABAQUS predictions and centrifuge test results

Group	Centrifuge test results			ABAQUS predictions			
	δ_0 (m)	P_{grp} (MN)	δ_{max} (m)	P_c (MN)	$r_{u,fail}$	δ_A (m)	δ_B (m)
			14.4m long piles				
G1	0.064	22.3	—	108.5 – 29.2	—	—	—
G2	0.056	25.5	0.11	111.3 – 29.6	—	—	—
G3	0.088	28.7	0.14	100.3 – 28.2	0.98	0.579	0.729
G4	0.120	31.9	0.40 [6.40]	90.5 – 26.7	0.57	0.473	0.606
			8.0m long piles				
G5	0.056	28.7	0.12	48.0 – 33.1	—	—	—
G6	0.114	38.3	1.41	33.8 – 23.0	0.00	0.114	0.149

TABLE 4. Pile group failure in the centrifuge and as predicted by ABAQUS

Group	Centrifuge observation	ABAQUS prediction
G1	Stable	Stable
G2	Stable	Stable
G3	Stable	Marginal
G4	Collapsed	Collapsed
G5	Stable	Stable
G6	Collapsed	Collapsed

The computer modelling suggests that group G4 could have failed with only partial liquefaction of the layer ($r_{u,base} = 0.57$). In the centrifuge tests, the group suffered an amount of displacement (0.4m) of the order of δ_A as the soil layer liquefied, so it is likely that by the end of the earthquake the upper hinges had formed, but high dilation of the soil around the piles as they displaced arrested motion. To illustrate this point, the pore pressure response adjacent to the piles in group G4 during the centrifuge tests are shown in Figure 6(b) along with free field responses at similar depths for comparison in Figure 6(a). The dashed lines indicate the excess pore pressures at full liquefaction. The large amount of both transient and residual dilation around the failed part of the pile can clearly be seen by comparing PPT 10952 to 10944. Similar dilative behaviour has also been observed by other researchers, notably Haigh & Madabhushi (2005). Pore pressures at lower levels near the pile group (PPT 6788) follow the free field behaviour

(PPT 6668). The dilative behaviour proved unsustainable however due to the drainage of pore fluid and the soil failed at around 200s (post-earthquake).

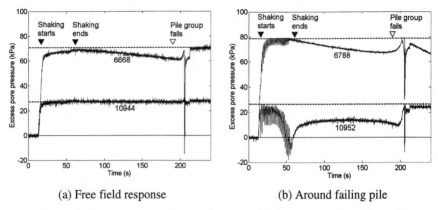

(a) Free field response   (b) Around failing pile

Figure 6: Pore pressure responses for group G4 (for locations, see Figure 1)

Finally, Table 3 suggests that group G6 should have collapsed prior to shaking. This was not observed to happen in the centrifuge tests and a closer look at the collapse loads reveals that the two loads are almost identical in magnitude. It is clear that observed deflections during shaking for this group exceeded both δ_A and δ_B, suggesting that G6 failed very soon after shaking started.

Failure of pile groups by instability can be related to initial imperfections in the pile group geometry and the depth to which liquefaction occurs. Larger imperfections can cause instability failures even with a shallow depth of liquefaction.

CONCLUSIONS

It has been demonstrated that axially loaded end-bearing pile groups can suffer instability failure during seismic liquefaction. The collapse load is a function of liquefaction propagation and initial imperfections in the geometry of the piles which may commonly occur during installation. Numerical modelling has suggested that even partial liquefaction can lead to collapse of imperfect pile groups with 80% of the total reduction in carrying capacity occurring within $0 < r_{u,base} < 0.4$ (i.e. 40% liquefaction at the base of the layer) for the uniform sand profile presented here. It has been shown that the finite element program ABAQUS can be used to undertake pseudostatic post-buckling analyses for pile groups modelled using the popular p-y methodology.

Acceptable predictions of collapse loads were obtained and validated against the results of six model pile groups tested in a geotechnical centrifuge. However, the numerical model is pseudostatic and therefore unable to account for the dynamic displacement-induced dilation in the soil around the piles, which leads to conservative predictions of collapse. These predictions were shown to out-perform estimates based on the elastic critical load at in the absence of soil support which over-predicted collapse loads.

REFERENCES

Baguelin, F., Frank, R. and Said, Y. H. (1977). "Theoretical study of lateral reaction mechanism of piles." *Géotechnique*, London, 27(3), 405-434.

Barton, Y. O. (1982). "Laterally loaded model piles in sand." *PhD Thesis*, University of Cambridge, UK.

Bhattacharya, S., Madabhushi, S. P. G. and Bolton, M. D. (2004). "An alternative mechanism of pile failure in liquefiable deposits during earthquakes." *Géotechnique*, London, 54(3), 203-213.

Boulanger, R. W., Kutter, B. L., Brandenberg, S. J., Singh, P., and Chang, D. (2003). "Pile foundations in liquefied and laterally spreading ground during earthquakes: centrifuge experiments & analyses." *Rep. No. UCD/CGM-03/01*, Dept. of Civ. & Envir. Engrg, University of California, Davis, CA.

Brennan, A. J. (2004). "Vertical drains as a countermeasure to earthquake-induced soil liquefaction." *PhD Thesis*, University of Cambridge, UK.

Crisfield, M. A. (1981). "A fast incremental/iteration solution procedure that handles 'snap-through'." *Computers and Structures*, Elsevier Science Ltd, 13, 55-62.

Fleming, W. G. K., Weltman, A. J., Randolph, M. F. and Elson, W. K. (1992). *Piling engineering*. John Wiley and Sons, N.Y.

Haigh, S. K., and Madabhushi, S. P. G. (2005). "The effects of pile flexibility on pile-loading in laterally spreading slopes." *Workshop on simulation and performance-based design of pile foundations in liquefied & laterally spreading ground*, University of California, Davis, CA, March 16-18.

Hardin, B. O., and Drnevich, V. P. (1972). "Shear modulus and damping in soils: design equations and curves." *J. of Soil Mech. And Found. Div.*, ASCE, 98(7), 667-692.

Hoadley, P. J. (1974). "Pile stability part 2 – non-elastic soil." *Proc. Conf. on Analysis and Design in Geotechnical Engineering*, ASCE, N.Y., 245-267.

Madabhushi, S. P. G., Schofield, A. N., and Lesley, S. (1998). "A new stored angular momentum earthquake actuator." *Proc., Centrifuge '98*, T. Kimura, O. Kusakabe and J. Takemura, eds., Balkema, Rotterdam, 111-116.

Maheetharan, A. (1990). "Modelling the seismic response of piles and pile groups." *PhD Thesis*, University of Cambridge, UK.

Schofield, A.N. (1981). "Dynamic and earthquake geotechnical centrifuge modelling." *Proc. Int. Conf. on Recent Advances in Geotechnical Earthquake Engineering and Soil Dynamics*, S. Prakash ed., St. Louis, USA, 3, 1081-1100.

HORIZONTAL-VERTICAL TWO DIMENSIONAL SHAKER IN A CENTRIFUGE

Jiro Takemura[1] and Jun Izawa[1]

ABSTRACT

This paper describes horizontal and vertical 2D shaker that can make horizontal and vertical motions of the shaking table independently in the TITech Mark III Centrifuge. In the 2D shaker, two inclined servo-hydraulic actuators with horizontal angle of 30 degrees are symmetrically mounted underneath the shaking table. Vertical static load of the moving mass is supported with four air springs so that the force from the actuators underneath the shaking table can act mainly for the dynamic motions of the mass. Results on a centrifuge tests using the shaker are also introduced. In the test effectiveness of active drainage as a countermeasure against liquefaction for a piled foundation was attempted to study.

INTRODUCTION

Since early 80's, there have been a lot of developments for shakers in centrifuge modeling (e.g., Arulanandan et al., 1982, Whitman, 1984) and now many types of shaker with various specification are available in many centrifuges all over the world, especially in Japan (Kimura, 2000). Shen et al. (1989) developed the bi-axial shaker, which can produce excitations in two horizontal directions. However, the shaking directions of all the shakers currently available are limited in the horizontal plane and no shaker that can simulate the vertical vibration has been developed. Vertical motion is sometimes dominated in P-wave which arrives earlier than the strong vibration, S-wave.

---

[1] Tokyo Institute of Technology, Tokyo, Japan.

Therefore the arrival of P-wave could be used to trigger an active control of seismic performance of structures. Acceleration of drainage from the sand is one of possible active type control against liquefaction.

In Tokyo Institute of Technology, an attempt has been made to develop a 2D shaker that can create horizontal and vertical motions of shaking table independently in the TITech Mark III Centrifuge (Takemura et al., 1999). In this paper some details of the 2D shaker and performance of the shaker in preliminary tests are first described. In order to discuss the effectiveness of active drainage as a countermeasure against liquefaction for a piled foundation, a shaking test of saturated loose sand with a piled foundation was also conducted using the shaker. In the test, vertical motion dominant preshock with small acceleration was first applied and then horizontal motion dominant mainshock with large acceleration followed the preshock. At the onset of the preshock, downward seepage was commenced by opening drainage valve at the bottom layer.

2D SHAKING TABLE

Preliminary consideration on 2D shaker

In order to create a table motion with two degrees of freedom, i.e. horizontal and vertical motions, at least two actuators are necessary for the shaker. Two types of shaking table were firstly figured to satisfy the minimum requirement for the horizontal-vertical 2D shaker as shown in Figure 1. One is the shaker with two actuators directed horizontally and vertically respectively and the other is the shaker with a pair of inclined actuators underneath the shaking table. In the former type, the horizontal and vertical actuators control the displacements of the table in the two ways independently. While in the latter type, the vertical and horizontal motions are controlled by the displacements of two actuators. Because of several reasons (Takemura et al.,2002), the latter arrangement with two inclined actuators was selected for the 2D shaker.

With the displacements of the left and right actuators, ΔL_L and ΔL_R, the shaking table moves ΔH and ΔV in the horizontal and vertical directions as shown in Figure 2(a). In order to create horizontal motion, ΔL_L and ΔL_R should be in opposite phase. If the displacements of the actuators are much smaller than the length of linkage (L) between the actuator and the shaking table, the horizontal motion without vertical one can be produced by putting ΔL_L and ΔL_R with the same amplitude as shown in Figure2 (b). The horizontal displacement of the table ΔH caused by $\Delta L_L(=-\Delta L_R)$ is given by

$$\Delta H = \Delta L_L \Big/ \cos\theta = -\Delta L_R \Big/ \cos\theta \qquad (1)$$

where θ is the horizontal angle of the inclined actuators. The vertical motion without

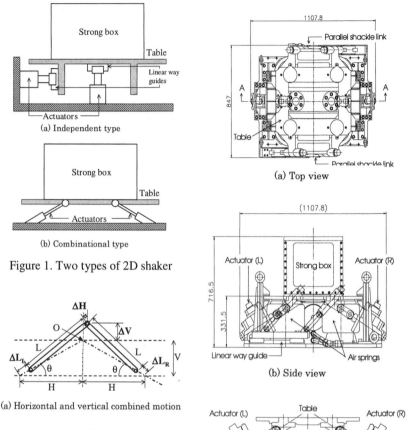

(a) Independent type

(b) Combinational type

Figure 1. Two types of 2D shaker

(a) Horizontal and vertical combined motion

(b) Horizontal motion

Figure 2. Horizontal and vertical motions caused by displacement of inclined actuators

(a) Top view

(b) Side view

(c) Sectional view at A-A

Figure 3. Schematic drawing of 2D shaker

horizontal one can be created by putting ΔL_L and ΔL_R with the same amplitude and phase. The vertical displacement of the table ΔV caused by $\Delta L_L(=\Delta L_R)$ is given by

$$\Delta V = \frac{\Delta L_L}{\sin \theta} = \frac{\Delta L_R}{\sin \theta} \qquad (2)$$

In order to control the horizontal and vertical components of displacement, ΔH and ΔV, simultaneously, the sum of the ΔL_L calculated from eqs.(1) and (2) and the sum of ΔL_R are input to the left and right actuators respectively.

The main difference between normal horizontal 1D shaker and horizontal-vertical 2D shaker is that the shaking table cannot be directly supported by liner way guides fixed tightly to the reaction frame. The linear way guides in the 1D shaker has two main functions, i.e. supporting the static vertical weight of the shaking table and model under centrifugal acceleration and preventing rocking motions of the shaking table. For the substitute of the former function of the linear way guide, the static weight is supported with four air springs so that the force from the actuators can be mainly utilized for the dynamic motions of the shaking mass. Furthermore, at the front and back side of the shaking table a pair of parallelogram shape shackle links were introduced between the shaking table and the liner way guides fixed to the reaction frame to mechanically prevent the rocking motion of the table as the latter function of the linear way guide in the 1D shaker.

System Description

Targeted specifications of the shaker are given in Table 1. The essential parts of the horizontal-vertical 2D shaker are two inclined servo-hydraulic actuators symmetrically mounted with horizontal angle of $\theta = 30°$, four air springs supporting the table and a pair of parallelogram shape shackle links between the shaking table and liner way guides on the reaction frame. Schematic drawing of the 2D shaker based on the considerations mentioned in the previous section is shown in Figure.3.

The linear hydraulic actuators attached on the reaction frame have a stroke of $\pm$ 3mm. Each actuator has force capacities of 77kN and 51kN for static and dynamic loading conditions with 20.5MPa oil pressure. Two servo-valves have been used for the actuators, MOOG J076-907 with higher frequency performance (200Hz, 90° phase delay) and lower flow rate (30L/mim at 7MPa pressure drop), and Tokyo Precision Instruments 275F with lower frequency performance (100Hz, 90° phase delay) and higher flow rate (75L/mim at 14MPa pressure drop). As the length of linkage between the actuator and the shaking table (L=81mm) is much larger than the possible actuator displacement (1-2mm), the horizontal and vertical displacements of the shaking table can be reasonably evaluated by eqs. (1) and (2). With the horizontal angle $\theta = 30°$, the vertical displacement ΔV is about 1.7 times as large as the horizontal displacement ΔH when the same displacements ΔL are given to the actuators. Hence with this condition as

Table 1. Targeted specification of 2D shaker.

Max. operational centrifugal acceleration	50g
Max. shaking mass including table	200kg
Table size	W640mm x B500mm
Stroke of table in horizontal direction	± 2.6mm
Stroke of table in vertical direction	± 1.5mm
Max. horizontal table acceleration	20g
Max. vertical table acceleration	10g
Max. velocity of actuator	220mm/s
Max. controllable frequency	200Hz

well as the maximum velocity of the actuator rod, in lower frequency range (up to 100Hz) the velocity of the actuator is critical in the performance of actuator and applicable acceleration in the vertical direction might be greater than that of horizontal one. However, in higher frequency range, the force of actuator becomes the critical condition and the applicable horizontal acceleration might be larger than the vertical one.

Details in the development of the shaker are given by Takemura et al.(2002).

PRELIMINARY TEST ON PERFORMANCE OF THE 2D SHAKER

Proof tests under loaded condition were carried out at 50*g* to confirm that the horizontal and vertical motions could be independently controlled using the 2D shaker with the MOOG servo valves. Horizontal and vertical motions were respectively applied to the table with frequencies of 50 and 100 Hz and input displacement amplitude of ± 1mm. The strong box filled with sand was put on the table. On the strong box additional mass was fixed to the proper position in order that the forces from the actuators passed the center of gravity of the shaking mass. This adjustment of the center of gravity is very important to avoid unfavorable rocking motions of the table. The mass of the box and sand used in the test was about 60kg.

Figures 4 show the observed time histories of horizontal and vertical accelerations at the center of the table for all the cases. Although the intended input displacement amplitudes were the same in all cases, i.e., ±1mm, in the cases of the frequency of 50Hz, the maximum vertical acceleration of 10g and the maximum horizontal acceleration of 6g were observed for the vertical and horizontal shaking tests respectively. On the other hand in the cases of 100Hz, the maximum accelerations were 7g and 20g for the vertical and horizontal shaking respectively. These results agree with the characteristic of the shaker caused by the actuator inclination angle and the maximum velocity of the actuator explained in the previous section. As one of the possible measures to improve

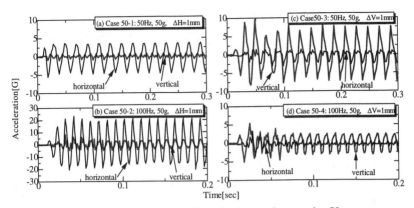

Figure 4. Acceleration time histories in proof test under 50g

the performance, a servo valve with higher flow rate (Tokyo Precision Instruments 275F) was used to create higher horizontal acceleration with larger velocity at the lower frequency range. The very large acceleration in the 100 Hz horizontal motions is probably because the frequency was very close to the natural frequency of the shaker in the horizontal motion in this particular case.

In the direction to which the motions was controlled to be zero, some acceleration was observed. However, the observed accelerations in this direction were in the range of 1/5 to 1/4 of those in the shaking directions, except at the beginning of the vertical shaking at 100Hz.

CENTRIFUGE TEST ON SAND LIQUEFACTION WITH ACTIVE DRAINAGE

Test Procedures and Conditions

In order to discuss the effect of active drainage on liquefaction of sand, a shaking test of saturated loose sand with a piled foundation was conducted using the shaker. In the test, an aluminum model container with inner dimensions of 450mm in length, 150mm in width and 270mm in height was used. 10mm thick rubber sheets were placed at the both side of the container to absorbed stress waves from the side boundary. Coarse silica sand No.3 was first compacted at the bottom of the container and then dry fine silica sand No.8 was placed by air pluviation in a very loose condition (ρ_d=1.27g/cm^3, Dr=30%). During the model preparation, various sensors were placed at the locations shown in Figure 5(a). The properties of the sand used are given in Table 2. After preparing the sand, the model container was placed in a vacuum tank and the san was saturated with water under vacuum pressure less than -90kPa.

Table 2. Material properties of soils used

	Fine silica sand No.8	Coarse silica sand No.3
Specific gravity: Gs	2.65	2.56
Mean particle size: D_{50}	0.10mm	1.47mm
Particle size: D_{10}	0.041mm	1.21mm
Uniformity coefficient: Uc	2.93	1.26
Max void ratio	1.333	0.971
Min void ratio	0.703	0.702
Hydraulic conductivity: k	2×10^{-5}m/s	5×10^{-3}m/s
k in 50G model	1×10^{-3}m/s	2.5×10^{-1}m/s

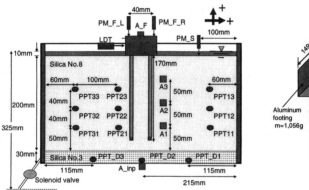

Figure 5. Test setup and location of sensors   Figure 6. Model piled foundation

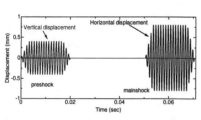

Figure 7. Input displacement to the shaker

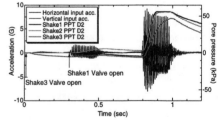

Figure 8. Input base motion, timing of opening solenoid valve and observed excess pore pressure at the bottom coarse sand layer

After saturating the model sand, the model piled foundation (Figure 6) was manually inserted into the ground and displacement sensors (PMs and LDT in Figure 5) were attached to the model. The container was then mounted on the shaking table and centrifugal acceleration was increased up to 50G. Confirming the equilibrium condition, the solenoid valve (CKD AB41-02-2) was opened to drain the water from the bottom coarse sand layer for about 1 second. The drainage rate depends on main factors, such as, the permeability and compressibility of the soil used, the head difference between the outlet and water level in the sand and the orifice size of the solenoid valve. After draining in the static condition, the shaking tests were conducted three times, Shake1, Shake2, and Shake3. In each shaking test, two vibrations were applied by sending signals equivalent to the displacement shown in Figure 7 to the controller of the shaker. In the first vibration called preshock and the second vibration called mainshock, vertical motion dominant sinusoidal waves with small acceleration and horizontal motion dominant sinusoidal waves with larger acceleration were aimed respectively. The input accelerations observed at the base of the container are shown in Figure 8. In the actual

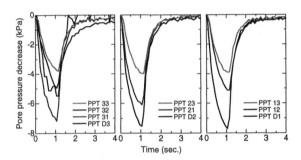

Figure 9. Pore pressure decrease by opening solenoid valve in static condition

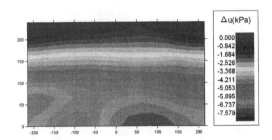

Figure 10. Contour of water pressure decrease by opening solenoid valve in static conditiono: t=1.07sec.

input waves, horizontal acceleration and vertical acceleration were observed in preshock and mainshock respectively. In Shake1 and Shake3, the solenoid valve in the drainage pipe connected to the bottom coarse sand layer was opened during the shaking, while in Shake2 the shaking test was conducted with undrained boundary at the bottom sand layer closing the valve. The timing of opening the valve is depicted in Figure 8.

Results and Discussions

Figures 9 and 10 show the decrease of pore pressure caused by opening of solenoid valve at the bottom of the model container in the static condition and the contour of the pore pressure decrease just before closing the valve respectively. In the test conditions, there was no significant variation of the pressure decrease in the horizontal direction and the pressure decreases of 5 - 6kPa were obtained at the deep part of fine sand layer, which is equivalent to about 10% of effective vertical stress.

Observed excess pore water pressures in the sand during the shaking tests are shown in Figure 11. PPT21 and PPT11 did not work for some reasons. Figure 12 shows

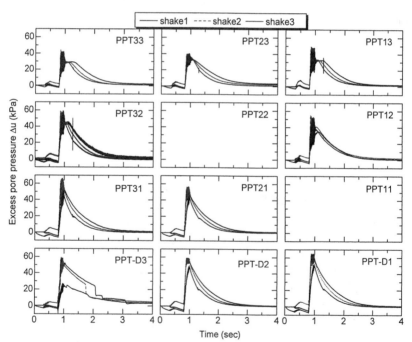

Figure 11. Observed excess pore pressures in each shake.

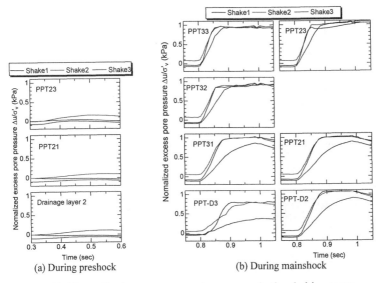

Figure 12. Normalized excess pore water pressure in the shaking tests:

(a) During preshock   (b) During mainshock

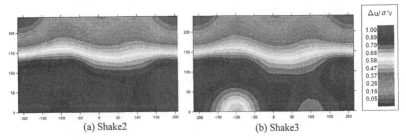

(a) Shake2   (b) Shake3

Figure 13. Contours of normalized excess pore water pressure in the shaking tests: t=1.0sec

the variations of normalized excess pore water pressure, $\Delta u/\sigma'_v$, in the preshock and mainshock for the three shaking tests. Contours of $\Delta u/\sigma'_v$, just after main shock of Shake2 and Shake3 are depicted in Figures 13. Although the value is not large, the effect of drainage from the bottom coarse sand, i.e., the difference of Δu for different drainage conditions can be observed in the preshock. However, this difference between Shake1 and Shake2 was ceased in the mainshock. It should be noted that the density of

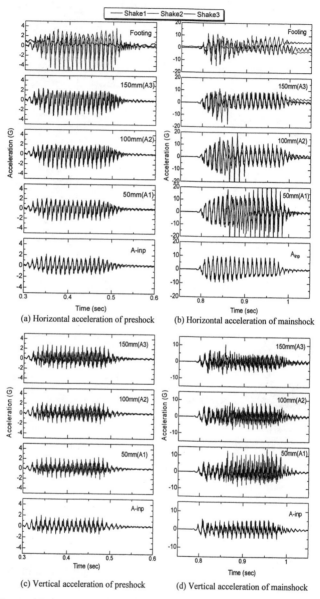

(a) Horizontal acceleration of preshock

(b) Horizontal acceleration of mainshock

(c) Vertical acceleration of preshock

(d) Vertical acceleration of mainshock

Figure 14. Observed accelerations during preshock and mainshock

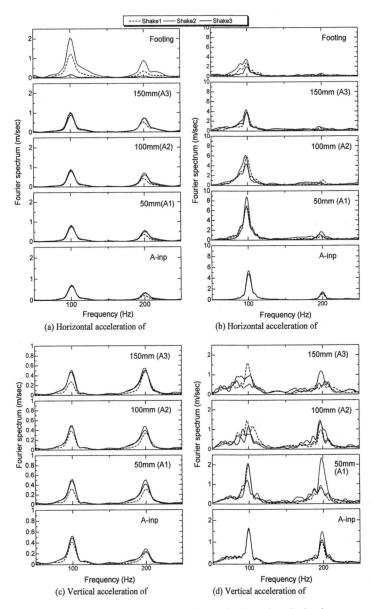

Figure 15. Fourier spectra in preshock and mainshock

the sand before Shake1 was smaller than that of Shake2. Because of the density difference, pore water dissipation after the mainshock is slower for Shake 1 than Shake2. Although this is also the case between Shake2 an Shake3, the rise of excess pore water pressure in the mainshock of Shake3 was much slower than that of Shake2, especially at deeper portion. This can be also confirmed in the contour of $\Delta u/\sigma'_v$ in Figure 13.

Horizontal and vertical accelerations observed by the vertical array of accelerometers in the ground and at the footing top are shown in Figures 14. Vertical accelerometer of the footing did not function. Fourier spectra of preshock waves and mainshock waves are also shown in Figures 15. The peak at 200Hz in the spectra can be attributed to rocking motion of shaking table and reflection of the waves from the boundaries. Although the input waves are different from the targeted one, it can be confirmed from the two figures that input waves in the three shaking tests are almost identical. Variations of waves in the upward direction are different both between preshock and mainshock and between horizontal and vertical accelerations. In the preshock, there are small differences in the spectra of horizontal acceleration in the sand regardless of the difference of shaking test. Large amplification of the footing acceleration in preshock of Shake2 could be attributed to the natural frequency of the piled footing closed to input frequency. While in the mainshock, amplification at deep depth and attenuation at shallow depth for 100Hz and clear attenuation for 200Hz were observed, but there is no clear amplification of the footing acceleration. This explicitly implies the deterioration of stiffness of sand in the mainshock. In the vertical acceleration responses, severe disturbances in the waves are observed in the mainshock, but the attenuation of high frequency component was less than that of horizontal acceleration.

Observed settlements of the footing and ground surface during the three shaking tests are shown in Figure 16. Footing settled almost vertically. Because the piles were floated in the loose sand, the footing penetrated into the sand due to shaking. The settlement of the footing took place only during shaking, while the ground settlement continued for a while after shaking. The effect of drainage on the settlement cannot be confirmed in three shaking tests in one model with very low initial relative density.

CONCLUDING REMARKS

The horizontal-vertical 2D shaker in the centrifuge has been newly developed. Details of the system were described in this paper. In consequence of the discussions on the performance of the shaker in preliminary tests, it was confirmed that the horizontal and vertical motions could be independently controlled using the shaker in the centrifuge, although there remain many things to improve the performance. With this modification, the performance of the shaker will be examined for further conditions, for example, combined accelerations and random motions.

An attempt was made to discuss the effect of active drainage on liquefaction of sand using shaking tests of a saturated loose sand model with a piled foundation. Although some differences were observed in the pore pressure and acceleration behavior, clear effectiveness of active drainage from the bottom could not be confirmed in a centrifuge model with very loose sand. Further tests with different conditions, e.g., density, pile bearing conditions, input motions are required for examining the effect of the drainage. To enhance the effectiveness of the drainage on the stability of piles in liquefiable sand, additional structures, such as, sheet pile wall with drainage facility surrounding the piles can be considered, which might be an objective for further study.

REFERENCES

Arulandan,K., Canclini, J.and Anandarajah,A.(1982). "Simulation of earth motion in the centrifuge." *J. of Geot. Eng.* ASCE, Vol. 108, No. GT5, 730-742.

Kimura, T. (2000). "Development of geotechnical centrifuges in Japan." *Proc. of Centrifuge 98*, Vol.2, Tokyo, Balkema, 945-954.

Shen, C. K., Li, X. S., Ng, C. W. W., Van Laak, P. A., Kutter, B. L., Cappel, K. and Tauscher, R. C. (1998). "Development of a geotechnical centrifuge in Hong Kong." *Proc. of Centrifuge 98*, Vol.1, Tokyo, Balkema, 13-18.

Takemura, J., Kondoh, M., Esaki, T., Kouda, M. & Kusakabe, O. (1999). "Centrifuge model tests on double propped wall excavation in soft clay." *Soils and Foundation*, Vol. 39, No. 3, 75-87.

Takemura, J., Takahashi, A. & Aoki,T.(2002). "Development of horizontal-vertical 2D shaker in a centrifuge." *Proc. of ICPMG '02*, St John's, Balkema, 163-168.

Whitman, V.R. (1984). "Experiments with earth quake ground motion simulator." *Proc. of Symp. on the application of centrifuge modelling to geotechnical design*, Manchester, 281-299.

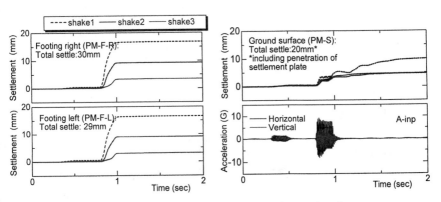

Figure 16. Settlement of piled footing and ground surface

SHAKING TABLE TESTS TO INVESTIGATE SOIL DESATURATION AS A LIQUEFACTION COUNTERMEASURE

Mitsu Okamura[1], Taiji Teraoka[2]

ABSTRACT

This paper discusses the possibility and effectiveness of soil desaturation as a liquefaction countermeasure technique. Soil type and soil characteristics that can be effectively desaturated during air injection, and longevity of air bubbles in the soil are described first. Inspection of samples taken by ground freezing method from SCP improved ground revealed that degree of saturation (S_r) shortly after SCP was lower than 91 % in the sand layers and a good correlation was found between S_r and 5% diameter of the soil. Degree of saturation of soils after several years ago was noticeably, but not significantly, higher as compared with that shortly after ground improvement, indicating longevity of air bubbles injected in the improved soil. Next, a series of shaking table tests of desaturated ground with and without rigid foundation resting on it is discussed. In the tests, vacuum pressure was introduced to the models to avoid distorted scaling factor.

INTRODUCTION

Along Japanese coast of Pacific ocean, huge earthquakes are expected to occur in the near future. In Kochi prefecture, extremely high Tsunami of 10m or higher is expected. Dikes protecting coastline in these area are often founded by liquefiable loose sand. Tsunami hazard map (Kochi Prefecture, 2004) indicates that if the dikes are completely slumped, most area in Kochi city, which is capital city of the prefecture, will be submerged but if the dikes don't subside at all, the flooded area

---

[1] Associate Professor, Ehime University, Matsuyama, Japan.
[2] Graduate Student, Ehime University, Matsuyama, Japan.

will be very limited. The remedial countermeasure against liquefaction is apparently needed for the dikes, but the local government almost gives up for doing any treatment for the dikes because of budget squeeze and of the ground improvement being very expensive. This is the case not only for Kochi prefecture but also many other local governments in Japan. Liquefaction countermeasure techniques that can be applied for existing structures are, in particular, costly. The above case, among others, clearly indicates strong needs to develop inexpensive liquefaction countermeasure techniques and desaturation by air injection is the possible one. This method is just to insert a pipe to liquefiable soils and exhaust air from the tip, as schematically depicted in Fig. 1.

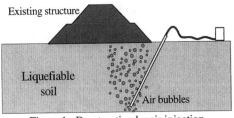

Figure 1. Desaturation by air injection

However, there is anxiety about stability of emitted air bubbles if they can be trapped and stayed long enough in the void of soils without dissolving in ground water and disappear eventually. Thus, soil type and soil characteristics that can be effectively desaturated during air injection, and longevity of air bubbles in the soil and a transition of degree of saturation over time are briefly described first. Then, a series of shaking table tests of desaturated ground with and without rigid foundation resting on it was discussed, aiming at studying effectiveness of desaturation as a liquefaction countermeasure.

DESATURATION OF FOUNDATION SOILS AND LONGEVITY OF AIR BUBBLES IN SOILS

The sand compaction pile (SCP) is widely used in Japan to improve loose soil profile. SCP is the ground improvement technique which densifies loose foundation soils by installing compacted sand piles. The compacted sand piles are constructed in such ways that a casing pipe is penetrated into ground to a depth and sand in the casing pipe is discharged into the bored hole during withdrawing the pile with an aid of pressurized air of the order of 500 kN/m^2 supplied from the top of the casing. It is observed during the sand pile construction that large amount of air which is exhausted with sand into the ground from the tip of the casing pipe continuously spouted from everywhere of the ground surface within the area about several meters from the casing. This is the common practice in construction of the sand compaction piles.   Of course, SCP is the ground improvement technique which does not desaturate soils intentionally.   But, inspection of such SCP improved soils reveals characteristics of desaturated soils by air injection.

Okamura et al. (2002) obtained high quality undisturbed samples using the ground freezing method at three sites and measured the degree of saturation.   The foundation soils of each site had been improved with SCP within a month before the samples were obtained.

Variation of degree of saturation with depth in the three sites shortly after SCP installation is given in Fig.2.   It is apparent that the improved soils in sand layers contained considerable amount of air.   The degree of saturation was lower than 91 % in the sand layers.   For clayey and silty soils, however, S_r was approximately 100 %, indicating a significant effect of grain size on the degree of saturation.   Then, degree of saturation was plotted against grain sizes, D_{10}, D_5, in Fig. 3 where D_{10} and D_5 denote 10 percent and 5 % diameter, respectively. S_r decreases with increasing D_{10} and D_5 with a relatively good correlation.

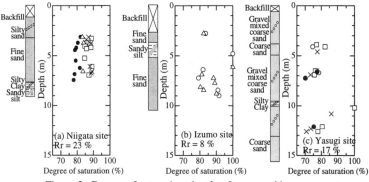

Figure 2. Degree of saturation shortly after ground improvement

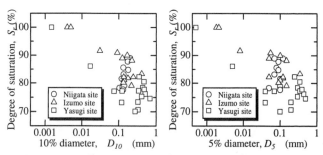

Figure 3. Variation of saturation degree with soil grain size

Next, in order to study the effect of time after ground improvement on the degree of saturation, they investigated another three sites in which foundation soils had been improved with SCP about 4 years, 8 years and 26 years ago, respectively (Okamura et al., 2005). It was revealed that air bubbles poured in the soil have survived for 26 years.

Results of laboratory tests summarized in the same manner as Fig. 3 indicated that the degree of saturation of improved soils more than several years ago seems to be noticeably, but not significantly, higher compared with soils shortly after the ground improvement, roughly at most 10 %. This fact implies that if degree of saturation after air injection is lower than 90 %, unsaturated condition lasts for a long time.

EFFECTS OF DESATURATION AS A LIQUEFACTION COUNTERMEASURE

In this section, effects of disaturation as liquefaction countermeasure are demonstrated through small scale 1g shaking tests. In modeling, all significant influences should be modeled in similarity, therefore, influential parameters on the behavior of desaturated soils under cyclic loading are selected and scaling lows are discussed.

Effect of Pore Fluid Stiffness and Initial Confining Stress

It has been recognized that degree of saturation (S_r) has significant effects on liquefaction resistance. Cyclic shear tests conducted in the previous studies indicated that even a several percent decrease in S_r doubles the liquefaction resistance (Yoshimi et al., 1989; Tsukamoto et al., 2002). Air in the void of soil plays a role of absorbing generated excess pore pressures by reducing its volume. Volumetric stiffness of pore fluid (air water mix) is one of dominant factors of liquefaction resistance. For a small change in pore pressure, Δp, volumetric strains of air and water are

$$\Delta p = \Delta \varepsilon_a B_a \tag{1}$$

$$\Delta p = \Delta \varepsilon_w B_w \tag{2}$$

and volumetric strain of fluid is

$$\frac{\Delta p}{B_f} = [(1 - S_r)\varepsilon_a + S_r \varepsilon_w] = \Delta p \left(\frac{1 - S_r}{B_a} + \frac{S_r}{B_w} \right) \tag{3}$$

where S_r is degree of saturation and B_a, B_w and B_f are bulk moduli of air, water and fluid, respectively. With the use of Boyle's law, and assuming the term S_r/B_w is zero, the modulus of fluid is given by

$$B_f = \frac{p}{(1 - S_r)} \tag{4}$$

where p is the absolute pressure of the fluid.

In order to absorb generated excess pore pressures, air bubbles have to contract. Volumetric strain of the bubble is approximately proportional to the generated pressure, the initial effective stress, σ_c', which is the possible maximum value of the excess pore pressure, is also a factor affecting liquefaction resistance.

Considering a model with a scale geometrically reduced by a factor of N, and with the prototype material being used, scaling factors are shown Table 1.   Parameters associated with compressibility of fluid are given and similarity conditions are expressed in terms of N-value.   It should be noted that scaling factor for absolute fluid pressure, $(p_0 + \gamma_w z)/(p_0 + N \gamma_w z)$, is nonlinear which varies with depth, from N at the ground surface to unity for the depth where hydrostatic pressure is much higher than atmospheric pressure, where p_0 is atmospheric pressure of 101 kPa and z is depth with reference to ground water table.   This may result in distortion of model tests results.   Consistent scaling factors can be attained by reducing the atmospheric pressure by 1/N.   Combination of the two parameters yields a dimensionless product, p/σ_c', which represents volumetric strain of fluid that can occur when the soil is liquefied, as indicated by equation (3).

Table 1. Scaring factors in scaled models

Parameter	Symbol	Dimensionless number	Scaling factor 1 atm.	1/N atm.
1 model length	l		1/N	1/N
2 soil density	ρ		1	1
3 acceleration	g		1	1
4 degree of saturation	S_r	S_r	1	1
5 absolute fluid pressure	p	$p/\rho g\, l$	N--1	1
6 initial effective stress	σ_c'	$\sigma_c'/\rho g\, l$	1	1

SHAKING TABLE TESTS

Two series of shaking table tests were carried out in this study. First one includes a series of tests on desaturated soils with level ground surface at different atmospheric pressures.   This was done to provide experimental evidence that the effects of desaturation on the liquefaction resistance were properly modeled.   In the latter series, on the other hand, effects of desaturation as a countermeasure for an existing structure were investigated.

Tests on Desaturated Soil with Level Ground Surface

Model preparation

The model was constructed in a container with internal dimensions of 90 cm length, 30 cm width and 60 cm height. A fully saturated loose Toyoura sand ($D_{50}=$

0.19 mm, G_s= 2.640, e_{max}= 0.973 and e_{min}= 0.609) bed was prepared in such a way that sand was liquefied and completely remolded by water jet spouted out from a nozzle. The nozzle was moved manually back and forth. A 5 mm diameter pipe was penetrated in the model and air was injected at locations so that air bubbles were uniformly distributed in the whole model. Degree of saturation was controlled at a the number of locations and time duration of air injection. The degree of saturation of the model was calculated from the change in the water table assuming the soil being fully saturated before the air injection. Then, a weak vibration was repeatedly applied to the model with an ample time interval until overall relative density of 70 % and depth of the bed of 330 mm were attained. The container was closed up tightly and a vacuum pressure was introduced to obtain lowered atmospheric pressure. The model was, then, shaken by a series of sinusoidal wave with a frequency of 5 Hz and peak accelerations (A_{max}) of either 0.33 g or 0.17 g. Excess pore pressure and acceleration were measured at locations shown in Fig. 4. Tests were conducted at different atmospheric pressures. Test conditions are summarized in Table 2.

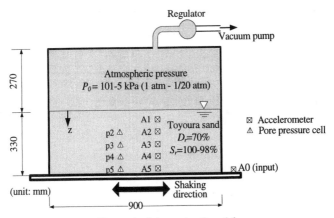

Figure 4. Schematic of model

Table 2. Test conditions of model with level ground surface

Relative density, D_r (%)	Degree of saturation, S_r (%)	Absolute atmospheric pressure, p_0 (kPa)	Max. input acceleration, A_{max} (g)	Scale factor, N	Num. of tests
64-66	100-98	6.5, 10, 18, 21, 31, 101	330	15.5, 10, 5.6, 4.8, 3.3, 1.0	16
65-66	100-99	11, 20, 40, 45, 65, 101	170	9.2, 5.0, 2.5, 2.2, 1.6, 1.0	11

Results and discussions

Figure 5 depicts comparisons of the excess pore pressure responses of saturated and desaturated models under atmospheric pressure of 101 kPa at three depths as well as the input acceleration time histories. Except for the duration of shaking, the input acceleration time histories of the two tests were quite similar, indicating remarkable reproducibility of the input motion. This was true through the course of the two series of tests conducted in this study. The excess pore pressures of the two models were also quite similar at any depth; both models liquefied almost simultaneously at around t = 3.6 second. This implies that degradation of saturation degree by 1 % has essentially no effect on soil liquefaction for the depth shallower than 0.3 m. This is probably due to the fact that pressure increase, Δp, was small and the volumetric strain of fluid given by equation (3) is not large enough to absorb the generated excess pore pressure.

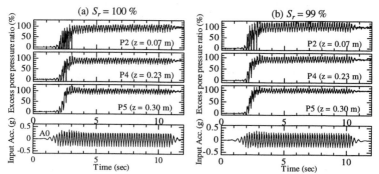

Figure 5. Excess pore pressure responses of saturated and desaturated models under atmospheric pressure of 101 kPa (A_{max} = 0.33g)

Figure 6 indicates the excess pore pressure time histories of desaturated soils at S_r= 99 %, subjected to base shaking with A_{max} = 0.33g, at different atmospheric pressures of 5, 10 and 31 kPa. Excess pore pressures for the three tests recorded at p2 (z = 0.07m) show little difference. Depths of p2 in the corresponding prototype are 0.21, 0.7 and 1.4 m. On the other hand, excess pore pressures at the depth of p4, of which corresponding prototype are 0.69, 2.3 and 4.6 m are apparently different; the lower the atmospheric pressure, the more number of cycles are required to reach liquefaction. This implies that degradation of saturation degree by 1% has a significant effect on soil liquefaction for the depths greater than about 1.4 m.

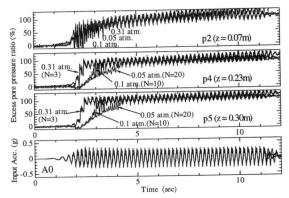

Figure 6. Excess pore pressure responses of desaturated models (S_r = 99 %) at different atmospheric pressure (A_{max} = 0.33 g)

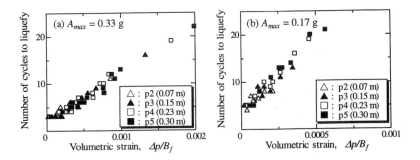

Figure 7. Relationship between volumetric strain and number of cycles to liquefy

As can be seen in equation (3), the volumetric strain, $\Delta p/B_f$, depends both on S_r and depth. The volumetric strain for each depth of the pore pressure cells of all the 16 tests for A_{max} = 0.33 g and 11 tests for A_{max} = 0.17 g were calculated. The number of cycles required to reach 100% excess pore pressure ratio obtained from each pressure cell records are plotted against $\Delta p/B_f$ in Fig. 7. For cases of A_{max} = 0.33 g, all the data points from the tests at different atmospheric pressure lie on a unique line. This is also the case for A_{max} = 0.17 g. These facts suggest that liquefaction behavior of desaturated foundation soil is properly modeled by reducing atmospheric pressure in accordance with scaling factor. It can also be seen in the figure that the number of cycles increases linearly with increasing volumetric strain up to 0.2%, with the slope being higher for smaller A_{max}.

Effects of Desaturation on Structure

Model Preparation and Test Conditions

In the second series of tests, a rigid acrylic block with outer dimensions of 150 mm high (H) and 150 mm wide (B) was placed on the ground surface, representing a rigid structure resting on the potentially liquefiable soil as shown in Fig. 8.   The load intensity at the base of the structure was 2 kPa, corresponding to a prototype stress of 20 kPa for models at atmospheric pressure $p_0 = 10$ kPa.

Table 3 summarizes test conditions.   All tests were conducted under the reduced atmospheric pressure of $p_0 = 10$ kPa, varying the width of desaturated zone; C1 is the model without a desaturated zone, the foundation soil was totally desaturated for C3, and the zone just beneath the structure was desaturated for C2.

The model was prepared in much the same way for the models C1 and C3.   While for model C2, in advance of air injection, a pair of cut-off walls (1 mm thick aluminum plate) was penetrated to the bottom of the model soil at an interval of 150 mm so that the injected air did not leak out of the zone.   The cut-off walls were extracted before the structure was set on the ground surface.   The models had approximately the same relative density of 65% and were subjected to the base shaking of $A_{max} = 0.4$g and 5 Hz frequency.

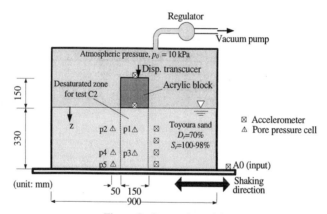

Figure 8. Setup of model

Table 3. Test conditions of model with structure

Test code	Relative density, D_r (%)	Degree of saturation, S_r (%)	Absolute atm. pressure, p_0 (kPa)	Width of desaturated zone (mm)	A_{max} (g)	Scale factor, N
C1	75	100	10	0	0.40	10
C2	78	98&100	10	150	0.40	10
C3	77	98	10	900	0.40	10

Results and discussion

Excess pore pressures observed in the tests C1 and C2 at two representative locations, that is the free field and below the structure, are compared in Fig. 9, together with settlement of the structure.   Note that excess pore pressure ratio was calculated assuming the load intensity owing to the structure as a uniform vertical surcharge.   For the fully saturated model C1, excess pore pressure ratio at free field (p2 and p4) attained 100% showing the soil liquefied in a few cycles.   Excess pore pressure below the structure (p1) increased at the beginning of the shaking event and started to decrease at $t = 3$ seconds, when the structure began to settle.   The decrease in the excess pore pressure below embankment is probably due to the large shear deformation of the soil.   This observation is consistent with that for shaking table tests of embankment models resting on liquefiable sand (Koga and Matsuo, 1990) as well as centrifuge tests (Adalier et al., 1998, Okamura and Matsuo, 2002).

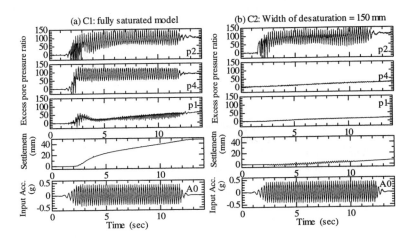

Figure 9.   Excess pore pressure responses of   models with different width of desaturated zone

On the other hand, excess pore pressure below the structure (p1), in which air was injected in the soil, was about half of that in model C1.   Although the depth of the pressure cell p2 was 0.07 m (0.7 m in the prototype sense), desaturation has an apparent effect to reduce excess pore pressure generation because of the effective overburden stress applied by the structure.   Settlement of the structure was reduced by a factor of 1/6.   It should also be noted in Fig. 9 that excess pore pressure at p4 was significantly lower than those in the model C1.   The cell p4 located in the untreated zone and 75 mm apart from the desaturated zone, which was influenced by the air in the desaturated zone.   Structure settlement at the end of shaking is depicted in Fig. 10.   Settlement of C3 was about half that of C2, indicating effect of width of desaturated zone on reducing settlement.

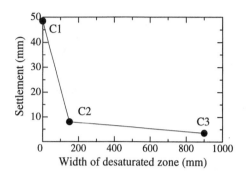

Figure 10. Settlement of structure at the end of shaking

CONCLUSIONS

It has long been known that degree of saturation has significant effects on liquefaction resistance of soils.   But, as far as authors are concerned, no one has devoted research efforts to develop a liquefaction countermeasure technique by soil desaturation.   This is probably because that most researchers and engineers think that air bubbles in soil voids may dissolve to ground water and disappear eventually. The recent research shows a clear possibility of the use of the soil desaturation as a liquefaction countermeasure technique.   Soil can be effectively desaturated during air injection and longevity of air bubbles in the soil are shown to some extent.

In the latter part of this paper, shaking table tests of desaturated ground with and without rigid foundation resting on it is discussed.   In the tests, vacuum pressure was introduced to the models to avoid distorted scaling factor.   Tests on models without foundation at different scaling factor, that is the model at different vacuum pressure, yielded   consistent   results   regarding   the   liquefaction   resistance,   confirming effectiveness of the vacuum technique.   It was also confirmed that volumetric strain of the voids due to seismic generation of excess pore pressure plays a dominant role in increasing liquefaction resistance of soil.   Desaturation is less effective for soils at shallower levels, say about 2m from ground surface.   In the tests on models with foundation resting on the ground surface, settlement was remarkably reduced by desaturation.

REFERENCES

Adalier, K., Elgamal, A.-W. and Martin, G. R. (1998). "Foundation liquefaction countermeasures for earth embankment, "J. of Geotechnical and Geoenvironmental Engineering, ASCE, 124(6), pp. 500-517.

Kochi Prefecture (2004). http://www.pref.kochi.jp/~shoubou/kochi_index/now.html

Koga, Y. and Matsuo, O. (1990). "Shaking table tests of embankments resting on liquefiable sandy ground, "Soils Found., 30(4), 162-174.

Okamura, M. and Matsuo, O. (2002). "Effects of remedial measures for mitigating embankment settlement due to foundation liquefaction." Int. J. Physical Modelling in Geotechnics, 2(2), 1-12.

Okamura, M., Ishihara, M. and Oshita, T. (2002). "Liquefaction resistance of sand deposit improved with sand compaction piles, "Soils Found., 43(5), 175-187.

Okamura, M., Ishihara, M. and Tamura, K. (2005). "Prediction Method for Liquefaction-induced Settlement of Embankment with Remedial Measure by Deep Mixing Method, "Soils Found., 44(4), 53-65.

Tsukamoto, Y., Ishihara, K., Nakazawa, H., Kamada, K., and Huang, Y. (2002). "Resistance of partially saturated sand to liquefaction with reference to longitudinal and shear wave velocities." Soils Found., 42(6), 93-104.

Yoshimi, Y., Tanaka, K., and Tokimatsu, K. (1989)." Liquefaction resistance of a partially saturated sand." Soils Found., 29(3), 157-162.

ANALYSIS OF GROUP PILE BEHAVIOR UNDER LATERAL SPREADING

Tetsuo Tobita[1] Member, ASCE, Susumu Iai[1],
Mikio Sugaya[2], and Hidehisa Kaneko[2]

ABSTRACT

Results of a series of centrifuge experiments to study the dynamic response of pile foundations under lateral spreading were compared to the results of numerical analysis. Experiments were carried out under the centrifugal acceleration of 40 G. Piles in the model foundation were lined up 3 by 3 pattern with a spacing of three pile diameter, and both pile head and bottom were rotation fixed. Piles were placed in the inclined ground of saturated sands and applied lateral loads due to the ground deformation. The effective stress finite element analysis was conducted to simulate these experiments. Computed time histories of pile head acceleration and displacement were consistent with those obtained from experiments. However, much smaller surface ground deformation in simulation might cause small amplitude of bending moments. Some calibration in the numerical modeling may be required to have more consistent results on bending moments.

INTRODUCTION

Many waterfront structures are built on deep foundations which are vulnerable to lateral loads especially due to the ground deformation during and after large earthquakes. Liquefaction-induced lateral spreading is one of the major causes of that ground deformation. (e.g., Mizuno 1987; Matsui and Oda 1996; Hamada and O'Rourke 1992; O'Rourke and Hamada 1992; Tokimatsu and Asaka 1998). Pile foundations not only support the inertial loads of the superstructures but also suffer the lateral loads due to the ground deformation. Therefore, the dynamic behavior of

---

[1] Disaster Prevention Research Institute, Kyoto University, Japan
[2] Kinki Regional Development Bureau, Ministry of Land, Infrastructure and Transport, Japan

soil-pile system have been intensively studied for the last decade. Dynamic behavior of pile foundations is highly nonlinear and influenced by numbers of parameters, such as material of a pile and soil, pile diameter and spacing, natural period of superstructures and soil-pile system. To study such a complicated phenomena, full scale experiment, although cases are limited, have been conducted (e.g., Brown et al. 1988; Peterson and Rollins 1996; Ashford and Rollins 2002). Brown et al. (1987) conducted large-scale tests for group pile subjected lateral load and proposed the p-multiplier concept to reshape the p-y curve of single pile to take into account the group effect. Rollins et al. (2005ab) conducted lateral loading tests with a full-scale pile group under blast-induced liquefaction and developed p-y curve for liquefied ground.

Instead of carrying out those full scale experiments, numerous small scale model tests have been employed using shaking tables or the geotechnical centrifuge (e.g. Tobita et al. 2004). Abdoun et al. (2003), for example, conducted centrifuge experiments with a slightly inclined laminar box to study pile response under lateral spreads of 0.7 to 0.9 m. One of their results is consistent with the observation that the maximum permanent bending moments occurred at the boundaries between liquefied and non-liquefied ground.

Numerical simulation is also widely used to simulate the dynamic response of pile foundations (e.g. Reese et al. 1996; Kitade et al. 2004). Most of two dimensional model uses p-y curve concept, or soil-pile interaction spring, that is required to simulate the three dimensional effects, such as slippage of soils near a pile surface. Therefore, it is important to properly determine the soil-pile interaction spring for a reasonable estimate of pile response. In the present study, the interaction spring obtained by the method proposed by Ozutsumi et al. (2003) is used to simulate those afore mentioned three dimensional effects observed in centrifuge experiments, and that is of prime objective of this paper.

CENTRIFUGE MODELING FOR GROUP PILE BEHAVIOR UNDER LATERAL SPREADING

Experiments were carried out with the geotechnical centrifuge at the Disaster Prevention Research Institute, Kyoto University (DPRI-KU). The centrifuge with rotation radius of 2.5 m has dual swing platforms at both ends of arms. The maximum capacity is 24 G-tons with a maximum centrifugal acceleration of 200 G. A shake table unidirectionally driven by a servo hydraulic actuator is attached to a platform and it is controlled through a personal computer (PC) on the centrifuge arm. All the equipment necessary for shake table control is put together on the arm. The PC is accessible during flight from a PC in the control room through wireless LAN and "Remote Desktop Environment" of WindowsXP (Microsoft, 2003). Capacity of the shake table is 15 kN, 10G and ±5 mm in maximum force, acceleration and displacement, respectively. Base excitation was given to a rigid soil box.

The uniform model ground was made of Soma-Silica sand No. 5 having the physical properties shown in Table 1 and the particle size distribution curve shown in Fig. 1. To study the behavior of pile foundations under lateral spreading, the model ground surface was inclined with the target inclination of 26.6° to the horizontal (vertical/horizontal=1/2) as shown in Fig. 2 and Photo 1, thus simulating rather steep slope which might be encountered in port facilities. The ground had flat part in downstream to simulate the sea floor. Sands were slowly sprinkled over the soil box filled with a viscous fluid. The target relative density of the ground was set to be 40 %. However, slightly lower values were obtained for all the cases as shown in Table 3. During the preparation of inclined ground, some fraction of surface soils were slowly moved downward due to the gravitational force, that might caused a development of the loose surface deposit.

A viscous fluid made of Metolose (Type: SM-25 Shin-Etsu Chemical Co.) was used to properly simulate pore water pressure dissipation during and after shaking. Metolose is water-soluble cellulose made of organic material. After preparing in the room temperature, the fluid was tested using a viscometer to achieve the specified viscosity (40cSt for 40g centrifugal acceleration) before pouring sands. There might be slight difference between the room temperature and the temperature in the centrifuge pit during flight but the effects of the difference was considered negligible because the duration of flight is about 20 to 30 min before the dynamic loads were applied.

Table 1. Soma-Silica sand No. 5.

e_{max}	e_{min}	D_{50}(mm)	U_c
1.105	0.685	0.38	1.5

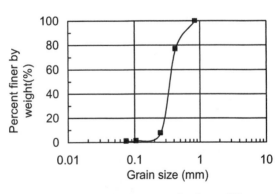

Figure 1: Particle size distribution curve for Soma-Silica sand No.5.

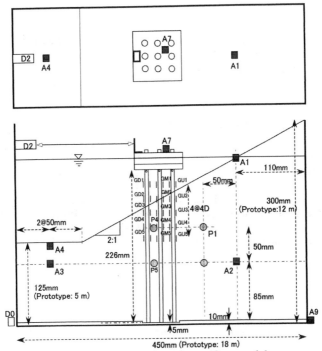

Figure 2: Cross section of centrifuge model.

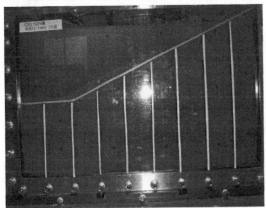

Photo 1: Side view of the model ground.

A pile with a 7 mm diameter brass tube simulated a prototype diameter pile of 0.28 m at 40 G. Other dimensions and mechanical properties of a model pile are shown in

Table 2. As shown in Fig. 2, group piles were lined up 3 by 3 with a spacing of 3 times a pile diameter. Both the pile top and bottom were set in rotation fixed condition. To achieve the fixity condition at the pile head, each pile in group was squeezed into a hole on an aluminum plate of 2 mm thickness, then a mass of 2.7 kg was placed on top and tied up with the plate. The weight of mass was determined so that the natural period of soil-pile system becomes about one second. The bottom of each pile was plugged into a hole on a bottom plate.

In a series of model tests, 6 accelerations, 2 displacements and 15 strain gage readings of a pile, and 4 pore water pressures were measured as shown in Fig. 2. A laser displacement sensor was used to measure pile head displacement. Accelerometers of strain gage type and pore pressure transducers of semi-conductor type were used. Total four cases are presented in this study as listed in Table 3, CS1: sinusoidal input with 40 Hz, CS2: synthesized near-field earthquake, CS3: synthesized plate boundary earthquake, and CS8: duplicated test of CS3. Input accelerations measured at the base of soil box are shown in Fig. 3. In what follows, units are all in prototype, unless otherwise noticed.

Table 2. Model pile properties.

	Model	Prototype	Unit
Length	0.25	10	m
Outer diamter (D)	7	280	mm
Wall thickness	0.9	36	mm
Young's modulus (E)	101	101	Gpa
Moment of inertia of area (I)	82	2.1×10^8	mm^4
Bending stiffness (EI)	8.2	2.1×10^8	$MN\text{-}mm^2$

Table 3. Experimental condition.

	Input	Dr (%)
CS1	Sin	35.4
CS2	Near Field	36.0
CS3	Plate Boundary	35.5
CS8	Plate Boundary	40.5

Ground Deformation After Shaking

Tracing the shape of the ground surface before and after shaking, and vertical markers attached on the inside of soil box, the ground deformation pattern is drawn as shown in Fig. 4. Large ground deformation of 2 to 4 m was occurred after shaking. It is clear that duration of shaking, i.e., 10, 60, and 100 seconds for CS1, CS2 and CS3/CS8, and the magnitude of surface displacement proportionally increase, and as duration becomes longer the ground surface tends to become flat. For CS1, compared with other cases the ground at deeper depth was mobilized and the surface displacement is about 3 m. While for CS2 and CS3/CS8, the surface displacements are, respectively, about 2 and 6 m, and the ground at shallower depth was mobilized. The pattern of the ground deformation of CS8 is consistent with the one of CS3.

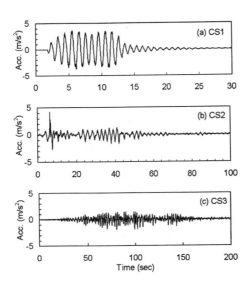

Figure 3: Input acceleration in centrifuge experiments.

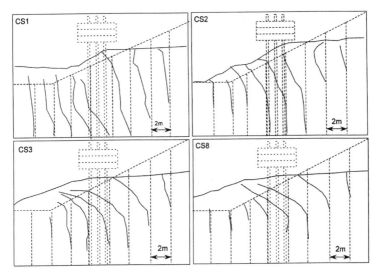

Figure 4: Ground deformation after shaking.

Pile Head Acceleration, Pile Head Displacement and Excess Pore Water Pressure

Results of a series of centrifuge experiments, CS1, CS2, and CS3, are shown in Fig. 5. Piles had no permanent deformation after experiments and therefore were within an elastic range. Initial vertical effective stress drawn in Fig. 5 is computed based on the vertical depth of the sensors (P1) from the ground surface. Comments for each case are as follows:

- CS1: Compared with the input acceleration shown in Fig. 3, the peak of measured pile head acceleration in Fig. 5 seems to be reduced with the development of excess pore water pressure. The pile head is shifted downward about 400 mm after about 5 seconds and the residual displacement of 300 mm is measured.
- CS2: Compared with the input acceleration shown in Fig. 3 (b), pile head acceleration is amplified about 100 %, even after the soil is liquefied. The pile head is moved downward with the maximum displacement of 600mm, and the residual displacement of about 80 mm is observed.
- CS3: The pile head acceleration is amplified similarly to CS2. Maximum displacement is 600 mm in downstream, and residual displacement is about 80 mm. Compared to the ground deformation of about 6 m as shown in Fig. 4, the residual displacement is small. This is because liquefied soil is soft enough to pass through the space between the piles.

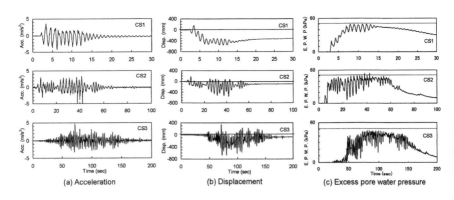

(a) Acceleration   (b) Displacement   (c) Excess pore water pressure

Figure 5: Time histories of pile head acceleration (a) and pile head displacement (b) and excess pore water pressure (c).

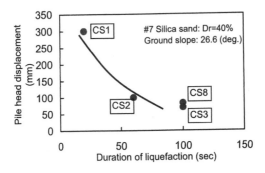

Figure 6: Duration of shaking and pile head displacement.

The relation between the duration of liquefaction and the pile head displacement obtained from a series of the experiments is summarized in Fig. 6. It is clearly shown that the residual pile head displacement becomes smaller as duration of liquefaction becomes longer, although, as mentioned earlier, the ground deformation contrary becomes larger. This is because liquefied soils are soft enough to pass through the space between piles. From the residual displacement of CS2 and CS3, the resistance force of pile and lateral force of the deformed ground may be in equilibrium at the pile head displacement of 80 mm.

NUMERICAL SIMULATION

Two dimensional effective stress finite element analysis, FLIP (Iai et al. 1992), is employed. Research is still on going and therefore results of CS2 only are presented. Physical properties of soil corresponding to the SPT value of 5 were assumed because of the lack of experimental results of the soil used in the centrifuge tests. Soil is modeled as having the multi-shear mechanism and piles are modeled with elastic beam elements. No joint elements but soil-pile interaction springs are implemented between soil and pile elements. The liquefaction strength curve in simulation is shown in Fig. 7. The meshes before and after the shaking are shown in Fig. 8. Compared to the one shown in Fig. 2, the pattern of ground deformation is similar but the magnitude of deformation is computed small. Pile head acceleration and displacement are compared in Fig. 9. The shape and amplitudes of both acceleration and displacement are consistent with the results of the experiment, except high spikes on the measured acceleration, and the amplitude of displacement which is computed smaller than the measured when piles moved downward.

Time histories of excess pore water pressure of P1 and P2 shown in Fig. 10 agree with experiments, however, for those of P4 and P5, simulation gives build up of pressure about one to two seconds earlier. Simulated pore pressure adjacent to piles may depend strongly on the response of piles. The elements to plot P4 and P5 are adjacent to the pile and therefore they may be subject to large shear compared to the elements of P1 and P4 surrounded by soil elements.

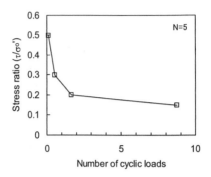

Figure 7: Liquefaction strength curve.

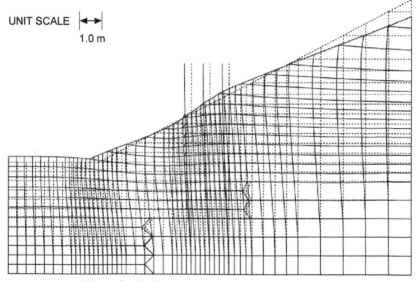

Figure 8: Mesh and deformation after shaking: CS2.

Time histories of bending moment for the downstream pile (GD1, GD3, and GD5) are plotted in Fig. 11. Computed bending moments from GD1 and GD3 are consistent with measured in terms of their amplitude and phase. As shown in Fig. 2, however, those are located above the ground surface. For the bending moments located under ground, GD5, computation gives smaller amplitude after 8 seconds than the ones measured in the experiment, This may be attributed to the small ground deformation in simulation, and require some calibration in the numerical model. Bending moments in depth is plotted in Fig. 12 for the time of 5.6 sec (a) and 41.8 sec (b). In

Fig. 12(a) at t=5.6 sec, the trend of the curve is similar each other, however, at time 41.8 sec shown in Fig. 12(b) piles above the ground surface were mostly bent in computation.

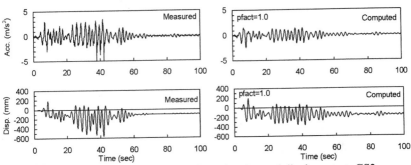

Figure 9: Comparison of pile head acceleration and displacement: CS2.

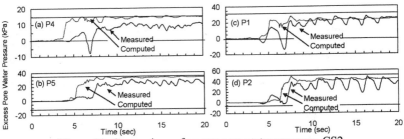

Figure 10: Comparison of excess pore water pressure: CS2.

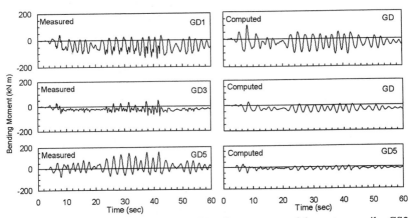

Figure 11: Comparison of time history of bending moment of downstream pile: CS2.

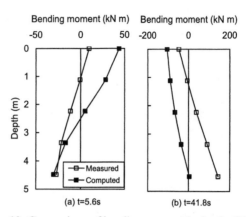

Figure 12: Comparison of bending moment in depth: CS2.

CONCLUSIONS

To study the dynamic response of pile foundations under liquefaction-induced lateral spreading, centrifuge experiments are conducted with inclined ground surface. Experimental results show that the duration of liquefaction is inversely proportional to the residual pile head displacements because liquefied soils are soft enough to flow between piles and give less lateral force and enough time for piles to be unloaded. This may possibly be observed in pile supported waterfront structures. However, the piles have to survive large lateral force that gives the maximum pile head displacement.

Numerical analysis based on the effective stress analysis, FLIP, properly simulated pile head acceleration, pile head displacement, and bending moment above the ground surface with a reasonable degree of accuracy. However, the ground deformation and the bending moments of piles under ground were simulated to be small compared to the one in the experiments. Further calibration is required numerical model.

REFERENCES

Abdoun, T., Dobry, R., O'Rourke, T.D., and Goh, S.H. (2003). "Pile response to lateral spreads: Centrifuge modeling." *Journal of Geotechnical and Environmental Engineering,* , ACSE, 129 (10), 869-878.

Ashford, S.A., and Rollins, K. M. (2002). "TILT: The Treasure Island liquefaction test." *Final Report No. SSRP 2001/17,* Department of Structural Engineering, University of California, San Diego.

Brown, D.A., Reese, L.C., and O'Neill, M.W. (1987). "Cyclic lateral loading of a large-scale pile group." *Journal of Geotechnical Engineering,* ASCE, 113 (11), 1326-1343.

Brown, D.A., Morrison, C., and Reese, L.C. (1988). "Lateral load behavior of pile group in sand." *Journal of Geotechnical Engineering,* ASCE, 114(11), 1261-1276.

Hamada, M., and O'Rourke, T.D. (1992). "Case histories of liquefaction and lifeline performance during past earthquakes." *Technical Report NCEER-92-0001*, National Center for Earthquake Engineering Research, State University of New York, Buffalo, 2 vols.

Iai, S., Matsunaga, Y., and Kameoka, T. (1992). "Strain space plasticity model for cyclic mobility." *Soils and Foundations*, Japanese Society of Soil Mechanics and Foundation Engineering, 32 (2), 1- 15.

Kitade, T., Kawamata, K., Ichii, K., and Iai, S. (2004). "Analysis of laterally loaded pile groups using 2-D FEM." *The 11th International Conference on Soil Dynamics & Earthquake Engineering, The 3rd International Conference on Earthquake Geotechnical Engineering*, 850-856.

Matsui, T., and Oda, K. (1996). "Foundation damage of structures." *Soils and Foundations*, Special Issue on Geotechnical Aspects of the January 17 1995 Hyogoken-Nambu Earthquake 189-200.

Microsoft. (2003). "Windows XP Professional." Microsoft Co., One Microsoft Way, Redmond, WA 98052-6399, USA.

Mizuno, H. (1987). "Pile damage during earthquake in Japan (1923-1983)." *Proc. Session on Dynamic Response of Pile Foundations, T. Nogami ed.*, ASCE, New York 55-77.

O'Rourke, T.D., and Hamada, M. (1992). "Case histories of liquefaction and lifeline performance during past earthquakes." *Technical Report NCEER-92-0001*, National Center for Earthquake Engineering Research, State University of New York, Buffalo, 2 vols.

Ozutsumi, O., Tamari, Y., Oka, Y., Ichii, K., Iai, S., and Umeki, Y. (2003). "Modeling of soil-pile interaction subjected to soil liquefaction in plane strain analysis." *Proceedings of the 38th Japan national conference on geotechnical engineering*, Akita, Japan 1899-1990.

Peterson, K.T., and Rollins, K.M. (1996). "Static and dynamic lateral load testing of a full-scale pile group in clay." *Civil Engineering Department Research Report, CEG. 96-02*, Brigham Young University, Provo, Utah.

Reese, L.C., Wang, S.T., Arrellaga, J.A., and Hendrix, J. (1996). "Computer program GROUP for windows, User's Manual, Version 4.0." Ensoft, Inc., Austin, Texas.

Rollins, K.M., Gerber, T.M., Lane, D.J., and Ashford, S.A. (2005b). "Lateral resistance of a full-scale pile group in liquefied sand." *Journal of Geotechnical and Environmental Engineering*, ASCE, 131(1), 115-125.

Rollins, K.M., Lane, D.J., and Gerber, T.M. (2005a). "Measured and computed lateral response of a pile group in sand." *Journal of Geotechnical and Environmental Engineering*, ASCE, 131(1), 103-114.

Tobita, T., Iai, S., and Rollins, K.M. (2004). "Group pile behavior under lateral loading in centrifuge model tests." *International Journal of Physical Modelling in Geotechnics, 4 (4)*, 1-11.

Tokimatsu, K., and Asaka, Y. (1998). "Effects of liquefaction-induced ground displacements on pile performance in the 1995 Hyogoken-Nambu earthquake." *Soils and Foundations*, Special Issue on Geotechnical Aspects of the January 17 1995 Hyogoken-Nambu Earthquake 163-177.

EVALUATING PILE PINNING EFFECTS ON ABUTMENTS OVER LIQUEFIED GROUND

Ross W. Boulanger, Member ASCE[1], Dongdong Chang[2],
Umit Gulerce[2], Scott J. Brandenberg, Member ASCE [2], and
Bruce L. Kutter, Member ASCE [1]

ABSTRACT

Earthquake-induced deformations of a bridge abutment underlain by liquefied soil may be reduced by the restraining forces that come from the pile foundation and bridge superstructure. The reduced abutment displacements may, in turn, reduce the loads or displacement demands that are imposed on the piles. Design methods that account for this "pile pinning" effect have been used in practice and included in the recent NCHRP 472 guidelines, but have not previously been evaluated against physical data. An initial evaluation of these pile pinning analysis methods is presented via comparison against the results of a recent dynamic centrifuge model test. Four aspects of these analysis methods are identified as requiring modification or adjustments to avoid an unconservative estimate of pile foundation performance. These and other design issues are discussed.

INTRODUCTION

Deformation of bridge abutments due to earthquake-induced liquefaction in the underlying foundation soils can impose large loads or displacement demands on pile foundations that are embedded through the abutments. The imposed loads and displacement demands depend on the extent to which the pile foundations and bridge

---

[1] Professor, Department of Civil & Environmental Engineering, University of California, Davis, CA, 95616.
[2] Graduate student researcher, Department of Civil & Environmental Engineering, University of California, Davis, CA, 95616.

superstructure may act to restrain the lateral displacements of the abutments. The final displacements of the both the pile foundation and abutments must be compatible, which requires that their interaction be accounted for in design. Analysis methods that account for this "pile pinning" interaction effect can reduce the expected foundation loads to values significantly smaller than those estimated without consideration of this pile pinning effect.

Design methods that account for this compatibility in displacement between the pile foundation and the abutment soils have been used in practice (Perez-Cobo and Abghari 1996, Law 2000, Zha 2004) and incorporated in the NCHRP 472 recommended specifications for seismic design of bridges (TRB 2002, Martin et al. 2002). These design methods can be summarized as consisting of the following three primary steps.

(1) Estimate the abutment displacement for a range of restraining forces from the piles and bridge superstructure. This step involves performing a slope stability analysis to estimate the yield acceleration, followed by a Newmark sliding block analysis to estimate the abutment displacement.

(2) Estimate the expected restraining force exerted on the abutment by the piles and bridge superstructure for a range of imposed abutment displacements. This step involves either a pseudo-static pile pushover analysis or some simpler approximation to determine the pile restraining forces or pushover curve.

(3) Determine the compatible displacement and interaction force between the abutment and the piles/bridge based on the intersection of the relations established in steps 1 and 2 above.

These design methods for pile pinning effects have not previously been evaluated or validated against physical data or well defined case histories.

This paper presents an initial evaluation of pile pinning analysis methods based on the results of a recently completed dynamic centrifuge model test. In the process of comparing computed and observed displacements, four primary issues were identified that need to be accounted for in design:

(1) The equivalent constant restraining force from the piles is smaller than the pile force at the end of shaking, which must be accounted for in determining the compatible displacement for the piles and abutment.

(2) The critical slide mass increases as the pile pinning force increases.

(3) The tributary slide mass width is greater than the abutment crest width.

(4) Deformations within the abutment can reduce the pile fixity above the liquefied layer, which reduces the shear resistance that the piles can provide.

The first of these issues is illustrated by the dynamic analysis of a rigid block, on an inclined plane, with a linear elastic spring restraining its displacement relative to the base. Then, a dynamic centrifuge model test of abutments with and without piles is described, and subsequently analyzed to illustrate the remaining three analysis issues listed above. Lastly, some general limitations and other practical considerations are discussed.

REVISED COMPATIBILITY CRITERION – SLIDING BLOCK EXAMPLE

Existing methods for evaluating pile pinning effects on bridge abutments (e.g. Martin et al. 2002, Zha 2004) determine the compatible displacement between the piles and abutment based on the intersection of a pile restraining force curve and an abutment displacement curve (described in more detail later). This approach is unconservative because: (1) the abutment displacement curve assumes that the pile restraining force is constant throughout shaking, (2) the pile restraining force actually begins at zero and increases throughout shaking as the piles are displaced, and (3) the compatibility criterion assumes that the final force in the piles acts as a constant restraining force on the abutment throughout shaking.

An example analysis is used to illustrate that the compatibility criterion needs to be determined using some measure of average pile restraining force. Consider a sliding block on an inclined plane, with the block also connected to a spring of stiffness K, as shown in Figure 1. The restraining force from the spring is $K\Delta$, where Δ is the displacement of the block relative to the base; Note that incremental sliding is limited to the down slope direction only in this example. The computed block displacement and spring force (normalized by block weight) versus time are shown in Figure 1, along with the imposed base acceleration (same motion as used for the centrifuge model). At the end of shaking, the block displacement is Δ_f and the spring force is $K\Delta_f$. If the spring was instead replaced by a constant restraining force in the sliding block analysis, what equivalent constant force would produce a reasonable estimate of the correct final displacement? If the equivalent constant force is taken to be $K\Delta_f$, then the computed sliding displacement will be significantly underestimated as shown in Figure 1. Alternatively, if the equivalent constant restraining force is taken to be $\frac{1}{2}K\Delta_f$, then the computed block displacement is reasonably close to the correct answer.

It is subsequently recommended that the equivalent constant restraining force from the spring (or piles) at a given final displacement be taken as the average restraining force that would develop as the displacement increased from zero to its final value. A curve that relates the equivalent constant restraining force to the final displacement can then be constructed as the running average of the pile restraining force versus displacement curve. For a linear spring, this would produce an equivalent constant restraining force equal to $\frac{1}{2}$ the final restraining force. For a rigid-plastic spring, this would produce an equivalent constant restraining force equal to the spring's yield strength. For an elastic-plastic spring, this would produce an equivalent constant restraining force curve that was: (i) equal to $\frac{1}{2}$ the elastic-plastic spring's restraining force up to the point of yield, (ii) approximately equal to the yield strength of the elastic-plastic spring at displacements that are many times larger than the yield displacement, and (iii) smoothly transitioning between the above two conditions at intermediate displacements. Initial parametric analyses using a range of input motions suggest that this approach is a reasonable approximation, given the uncertainties inherent to the overall procedure.

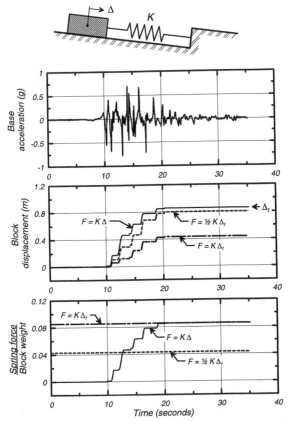

Figure 1. Simple block with restraining spring and its comparison to analysis results using equivalent constant restraining forces.

CENTRIFUGE MODEL

A centrifuge model of pile pinning effects on bridge abutments was recently completed on the 9-m radius centrifuge at UC Davis. A schematic cross-section of the first centrifuge model is shown in Figure 2. The soil profile consisted of loose Nevada sand ($D_r \approx 35\%$) overlying dense Nevada sand ($D_r \approx 75\%$). Thin layers of nonplastic silt were located above and below the loose sand layer. The water table was at the top of the upper silt layer. The two abutments were constructed of coarse Monterey sand, with the crest sloping slightly toward the channels, and with side slopes of 2:1 (horizontal:vertical) on all sides. The model was tested at a centrifugal acceleration of 60g, giving a prototype abutment height of about 8.4 m near the channel.

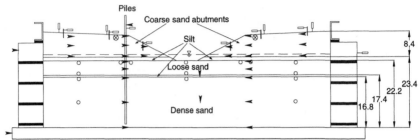

Figure 2. Cross-section of centrifuge model with two facing abutments
(Prototype dimensions in m).

A row of six piles were driven along the head of one abutment slope. The piles scaled to 0.72-m diameter in prototype, and were spaced four diameters apart along the 12-m wide abutment.

The test was shaken with a single earthquake motion having a peak base acceleration of about 0.78 g. The lateral displacement at the head of the abutment slopes was approximately 1.6 m for the abutment without piles and approximately 1.2 m for the abutment with piles. The photograph in Figure 3 shows the model during its dissection after testing.

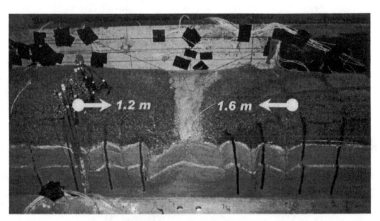

Figure 3. Centrifuge model during dissection.

SLOPE STABILITY AND YIELD ACCELERATION

Pile pinning analyses are now presented for the two abutments in the centrifuge model tests. These analyses illustrate the general methodology, as well as the effect of several important modifications to previously recommended procedures.

First, slope stability analyses are used to determine the yield acceleration for the abutment. The analyses presented herein used Spencer's method with noncircular slip surfaces. The critical slip surface was constrained to go through the middle of the loose sand layer that liquefied during earthquake shaking, and to intersect the piles at the middle of the layer, as shown in Figure 4. The residual shear strength of the liquefied sand was taken as 25 kPa because, as will be shown later, it produces computed displacements that are comparable to the observed displacements for the abutment without piles.

The critical slip surface with zero or small restraining force from the piles was located close to the face of the abutment slope. Once the pile restraining force exceeded a certain level, the critical slip surface quickly shifted further away from the abutment slope, such that the slide mass was greatly increased (Figure 4).

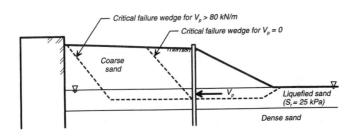

Figure 4. Critical slip surfaces with and without significant pile restraining force.

The design methodology described by Zha (2004) determines the critical slide mass for the case without piles, and then treats the pile pinning force as restraining only that slide mass. This approach neglects the increase in slide mass volume with increasing pile restraining force that is illustrated in Figure 4. The NCHRP 472 guidelines do not explicitly address this issue. The critical slip surface must be allowed to grow with increasing pile restraining force to avoid over-predicting the beneficial reduction in abutment displacement caused by pile pinning.

The computed yield acceleration for the slope is plotted versus the pile restraining force in Figure 5(a). This relation is shown for two cases. The first case is when the critical slip surface is determined for the case of zero pile restraining force, and then this critical slip surface is used for all subsequent calculations. The second case is when the critical slip surface is allowed to vary for different pile restraining forces.

These results show that the yield acceleration, for a given level of pile restraining force, is much smaller in the second case because a much larger mass of soil has to be restrained from sliding.

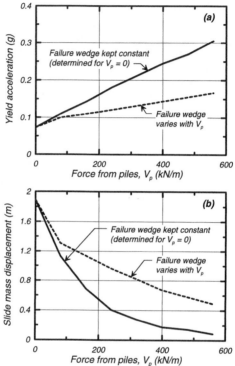

Figure 5. Yield acceleration and slide mass displacement versus pile restraining force.

SLIDING BLOCK CALCULATION OF ABUTMENT DISPLACEMENTS

Abutment displacements were then computed using a Newmark sliding block analysis with different amounts of pile restraining force. The pile restraining force was assumed to be constant throughout shaking, as is assumed in existing design methods (e.g., Zha 2004, Martin et al. 2002). The input motion was taken as one of the recorded acceleration time series from just below the loose sand layer (i.e., near the top of the dense sand layer). The computed abutment displacement is plotted versus pile restraining force in Figure 5(b) for the same two cases as used in Figure 5(a). As expected, for a given pile restraining force, the abutment displacements are much larger if the critical slide surface was allowed to vary with the applied pile restraining force.

PSEUDO-STATIC PILE PUSHOVER ANLAYSIS

A pseudo-static pushover analysis of the piles was performed using a beam on nonlinear Winkler foundation (BNWF) model. The analysis was performed using the FE platform OpenSees. The p-y springs were based on API recommendations for sand, and the effects of liquefaction were accounted for using procedures described in Boulanger et al. (2003). The computed shear force in a single pile was determined to be about 900 kN, as shown in Figure 6. This shear force is very close to the value estimated by assuming that plastic hinges form in the pile at three pile diameters (i.e., 3D) above and below the liquefied layer, and that the soil in between the plastic hinges applies negligible lateral loading (follows NCHRP 472, except that a distance of 5D was used to be more conservative). As indicated in the insert in Figure 6, this simple equation results in a computed pile shear resistance of about 880 kN, which is very close to that produced by the BNWF pushover analysis. If the plastic hinges were instead assumed to be at 5D above and below the liquefied layer, then the pile shear resistance would be estimated as 670 kN by this simple expression.

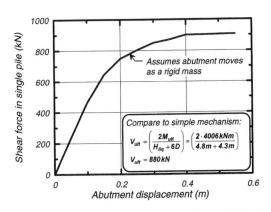

Figure 6. Pseudo-static pushover curve for a single pile.

DISPLACEMENT COMPATIBILITY BETWEEN ABUTMENT AND PILES

Effect of Compatibility Condition

The displacement compatibility between the abutment and piles was subsequently evaluated as shown in Figure 7. The pile resistance is shown for two cases: the pile resistance V_p versus a given abutment displacement (Δ_{abut}), and the displacement-averaged pile resistance (V_p)$_{ave}$ versus a given abutment displacement. In either case, the restraining force from the six piles are added together, and then divided by the width of the abutment crest to arrive at the equivalent resistance per m width of abutment (as described in NCHRP 472). The abutment displacements are then shown for the two cases previously described in Figure 5(b). The most optimistic solution would be an expected abutment displacement of about 0.2 m, based on the intersection of the V_p-Δ_{abut} pushover curve and the sliding block calculation with the critical slip surface kept constant (independent of V_p). The more realistic solution would be about 0.7 m, based on the intersection of the $(V_p)_{ave}$-Δ_{abut} curve and the sliding block calculation with the critical slip surface varying with the pile restraining force. However, both solutions under-estimate the measured abutment displacement of 1.2 m.

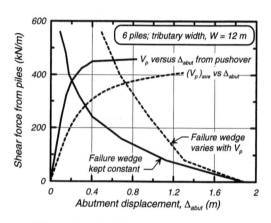

Figure 7. Compatibility points for different representations of the pile pushover results and the abutment displacement calculations.

Effect of Tributary Width

The tributary width of the abutment is often assumed to be equal to the abutment crest width (e.g., NCHRP 472), but in reality it will be greater due to the influence of the abutment side slope masses. Consider the cross-section of the centrifuge abutments shown in Figure 8. The piles were positioned across the full crest width of the abutment. As the abutment tries to displace toward the river channel, the piles will act to restrain a portion of the slide slope masses as well. Assuming that one-half of the slide slope masses (areas "a" and "c" in Figure 8) must be restrained by the piles, then the equivalent width of the abutment (for the same total height of 8.4 m) would actually be 20.4m as opposed to the crest width of 12 m.

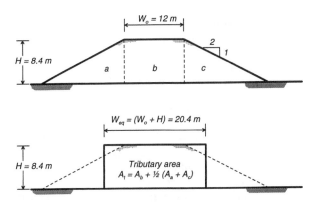

Figure 8. Equivalent tributary width when one-half the mass of the abutment side slopes are included in the tributary mass.

The displacement compatible solution was then determined using this wider tributary width of 20.4 m, with the results shown in Figure 9. The change in tributary width from 12 m to 20.4 m causes the compatibility point to increase from about 0.7 m to about 1.0 m.

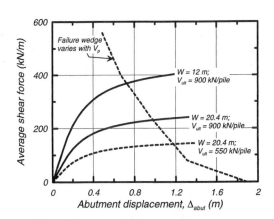

Figure 9. Compatibility when tributary width includes a portion of the side slope mass and pile resistance is reduced for loss of pile fixity in the deforming abutment.

Effect of Pile Fixity in the Abutment

The measured pile shear resistances in the centrifuge test were significantly lower than expected based on the pseudo-static pushover analysis. These shear resistances were measured using both shear strain gages and through differentiation of bending moment profiles. These data were still being processed at the time of this workshop, and so the results are not yet finalized. Nonetheless, it appears that the shear resistance in the piles would have been at most about 550 kN, which is slightly more than one-half the value predicted by the pushover analyses.

The lower shear resistance in the piles may be attributed to deformations within the abutment mass and in the deeper dense sand layer which reduced the bending fixity of the pile above and below the liquefied layer, thereby softening the piles' resistance to lateral loading from the abutment. The pushover analysis had assumed, as is common, that the abutment soil and deep dense sand layer would exhibit zero shear strain, with the abutment displacement being attributed entirely to shear strains in the liquefied layer. In the centrifuge model, however, deformations and shear strains developed throughout the abutment mass and in the dense sand layer, as illustrated by the colored sand markers exposed in the model excavation (Figure 3). For this model, the pushover analyses would have to include a more reasonable estimate of the displaced shape of the ground, including shear strains in the abutment mass and in the dense sand, to produce a more realistic estimate of the shear resistance from the piles. Similarly, the simple expression for pile shear resistance given in NCHRP (or the insert in Figure 6) will over-estimate pile shear resistance whenever the deformations of the abutment mass and deeper nonliquefied layers are significant enough to reduce pile bending fixity above and below the liquefied layer.

The effect of this lower pile shear resistance on the compatible pile and abutment displacements is illustrated in Figure 9, showing how reducing the pile shear resistance from 900 kN/pile to 550 kN/pile increases the compatible displacement from about 1.0 to 1.2 m. This final estimate of compatible displacements, taking into account the various issues previously discussed, is in reasonable agreement with the pile displacements observed in the centrifuge model.

DISCUSSION

Abutment deformations due to liquefaction in the foundation soils can involve complex shear and extensional strains, cracking of the abutment into discrete blocks that move independently of each other, out-of-plane slumping, and settlements due to reconsolidation of the liquefied soils. The liquefied soil's stress-strain behavior is highly nonlinear, with deformations developing as shear strains cyclically accumulate throughout the liquefied layer and with additional displacements sometimes developing along localizations that form due to void redistribution at the interfaces between liquefied layers and overlying low-permeability soils such as silt (e.g., Kulasingam et al. 2004). A Newmark analysis does not account for any of these mechanisms, and consequently has very limited predictive capabilities. Nonetheless, the method does provide a simple means to evaluate the sensitivity of the pile pinning solution to uncertainties in the various input parameters (e.g., residual shear strength, ground motion, pile shear resistance), and can be used to bracket the range of expected behaviors given the unavoidably large uncertainties in the input parameters.

The dynamic interactions between pile foundations and laterally spreading ground are only crudely approximated by the combination of a Newmark analysis for the abutment and a static (pushover) analysis for the piles. The actual dynamic interaction is highly complex (e.g., Brandenberg et al. 2005), which introduces additional uncertainties in the predicted responses. Nonetheless, these simplified analyses do provide a first-order approximation of the pile pinning effect, and can be useful for design purposes provided that the uncertainties in the input parameters and analysis method are reasonably accounted for. For example, a pile pinning analysis may indicate that lateral spreading displacements of an abutment will be limited, but the abutment backfill will still experience internal deformation (out of plane slumping, reconsolidation settlement, etc) that may affect serviceability of the bridge abutment approach after an earthquake.

SUMMARY AND FUTURE DIRECTIONS

An initial evaluation of pile pinning analysis methods for bridge abutments was presented along with the results of a recently completed dynamic centrifuge model test for comparison. The centrifuge model test showed slope displacements of about 1.6 m in an abutment without piles and about 1.2 m in an abutment with a row of six 0.72-m diameter piles. Current pile pinning analysis methods would have predicted much smaller pile displacements than were observed. The following modifications to these pile pinning analysis methods were introduced and subsequently shown to result in reasonable agreement between computed and observed pile displacements.

- The increase in the critical slide mass with increasing pile pinning force must to be explicitly accounted for.
- The equivalent "constant" restraining force from the piles should be taken as the average pile restraining force up to the final displacement.
- The tributary mass of the abutment should include a portion (e.g., ½) of the side slope masses, and not just the mass of the soils behind the crest width.
- The pile shear resistance across the liquefied layer must account for the potential reduction in pile fixity above or below the liquefied layer that can occur due to internal abutment deformations or shear strains in the underlying strata.

These pile pinning analysis methods involve several crude approximations that are expected to contribute to significant uncertainty in the predicted responses. Nonetheless, the methodology can still be useful for bracketing the range of likely responses given the inherent uncertainties in the input parameters and analysis results. Further evaluations of these analysis methods are required, and are the subject of ongoing studies involving centrifuge model testing and associated analyses.

ACKNOWLEDGMENTS

Funding was provided by Pacific Earthquake Engineering Research (PEER) Center, through the Earthquake Engineering Research Centers Program of the National Science Foundation, under contract 2312001, and the PEER Lifelines program under contract 65A0058. The contents of this paper do not necessarily represent a policy of either agency or endorsement by the state or federal government. Recent upgrades to the centrifuge have been funded by NSF award CMS-0086566 through the George E. Brown, Jr. Network for Earthquake Engineering Simulation (NEES).

REFERENCES

Boulanger, R. W., Kutter, B. L., Brandenberg, S. J., Singh, P., and Chang, D. (2003). "Pile foundations in liquefied and laterally spreading ground: Centrifuge experiments and analyses." Report No. UCD/CGM-03/01, Center for Geotechnical Modeling, Department of Civil & Environmental Engineering, University of California, Davis, 205 pp.

Brandenberg, S. J., Boulanger, R. W., Kutter, B. L., and Chang, D. (2005). "Behavior of pile foundations in laterally spreading ground during earthquakes." Journal of Geotechnical and Geoenvironmental Engineering, ASCE, in press.

Kulasingam, R., Malvick, E. J., Boulanger, R. W., Kutter, B. L. (2004). "Strength loss and localization at silt interlayers in slopes of liquefied sand." Journal of Geotechnical and Geoenvironmental Engineering, ASCE, 130(11), 1192-1202.

Law, H. (2000). "Lateral spreading on pile at the Oakland Mole SFOBB East Span Seismic Safety Project," Memorandum 09-12-2000, Earth Mechanics, Inc., Fountain Valley, California.

Martin, G. R., March, M. L., Anderson, D. G., Mayes, R. L, and Power, M. S. (2002). "Recommended design approach for liquefaction induced lateral spreads." Proc. 3rd National Seismic Conference and Workshop on Bridges and highways, MCEER-02-SP04, Buffalo, NY.

Perez-Cobo, A., and Abghari, A. (1996). "Lateral spreading and settlement potential for Salinas River Bridge (Bridge No. 44-0002)." Memorandum 02-22-1996, Caltrans, Office of Structural Foundations, Sacramento, California.

Transportation Research Board (2002). Comprehensive Specification for the Seismic Design of Bridges. National Cooperative Highway Research Program (NCHRP) Report 472, National Research Council, 47 pp.

Zha, Jin-xing (2004). "Lateral spreading forces on bridge abutment walls/piles." Geotechnical Engineering for Transportation Projects, Geotechnical Special Pub. 126, M. K. Yegian and E. Kavazanjian, eds, ASCE, Vol. 2, 1711-1720.

Subject Index

Page number refers to first page of paper

Author Index

Page number refers to first page of paper